C000172742

INSIDE THE COSMIC MIND

'Scholarship in the field of archetypal cosmology situates astrology in a rich lineage of ancient myth and philosophical speculation, and demonstrates its congruence with certain new-paradigm perspectives in modern science. Here Phoebe Wyss skilfully explores some major elements of this emerging vision, leading the reader from first principles to the practical application of astrology to individual biography. In a style both engaging and accessible, *Inside the Cosmic Mind* is a valuable and much-needed guide to the theory and practice of an archetypal approach to astrology.'

KEIRON LE GRICE, AUTHOR OF *THE ARCHETYPAL COSMOS*

~

'Building on the work of Richard Tarnas and Keiron Le Grice on archetypal astrology, this fascinating book reaffirms the centrality of the microcosm-macrocosm/individual-universal correspondences and weaves it in with the work of Jung, especially on synchronicity, to present a vision of oneness that can respond to the loss of meaning in modern times and the need to renew our culture. Phoebe also offers a detailed analysis of the life of William Blake in the light of his natal chart as well as explaining the history, background and significance of archetypal astrology.'

DAVID LORIMER, SCIENTIFIC AND MEDICAL NETWORK

~

'This ambitious and highly erudite book seeks to present us with a worldview which could lead towards a renewed spirituality in which astrology plays a central role. As with all astrological traditions it rests on the presumption that human life, and the living world as a whole, is closely associated with the timeless order of the cosmos. The author's attempt to bring astrology back into the mainstream of modern thought is an interesting and provocative one, drawing on a remarkably wide spectrum of traditions, and propelling the argument with eloquence.'

J.J. CLARKE, AUTHOR OF *THE SELF-CREATING UNIVERSE*

INSIDE THE COSMIC MIND

Archetypal Astrology and the New Cosmology

PHOEBE WYSS

Floris Books

To a new vision for the second half of the twenty-first century, and to Sylvain, Alia, May and Anouk who will go forward into it.

Published in 2014 by Floris Books
© 2014 Phoebe Wyss

Phoebe Wyss has asserted her rights under the Copyright, Designs and Patent Act 1988 to be identified as the Author of this Work

All rights reserved. No part of this publication may be reproduced without the prior permission of Floris Books, 15 Harrison Gardens, Edinburgh
www.florisbooks.co.uk

This book is also available as an eBook

British Library CIP Data available
ISBN 978-178250-110-7
Printed in Great Britain
by Bell & Bain, Glasgow

Contents

List of Figures

APPENDICES

Preface

This book is intended as a general introduction to the new paradigm of an archetypal cosmos and also as a justification of astrology. Astrology can be explained, according to the author, in the context of this newly emerging worldview.

In addition to offering a plausible explanation of how astrology works, the book also provides a guide and a map of the way for newcomers to astrology. The elements of an astrological chart are explained within their cosmological context, and a simple top-down method of chart interpretation is introduced.

The archetypal approach to astrology presented here offers an accessible entry into this complex subject. And further help is given to those needing it by the Glossary and Appendices, which can be referred to for the meanings of the technical terms used, and for diagrams and tables that help to explain astrology's basic concepts.

Introduction

Astrology is marginalised in our society. Astrologers are at the most tolerated as providers of trivial entertainment in horoscope columns. Some are accused of being charlatans; others are blamed for dabbling in the dark arts, while across the board astrology is labelled as 'unscientific'. Its opponents are vehement, convinced that those who believe in astrology must be either deceivers or deceived, because there is no way it can possibly work – or is there?

Like everyone else who has taken time out to study astrology in depth I know it delivers. I've been using it successfully for thirty-five years in my consultations – though I was never able to explain how it works, which often left me helplessly wrong-footed. It also planted a big question mark over my chosen vocation, that of an astrologer. So in 2005 I started out on a quest to find the answer to this mystery, and this book is the record of the path I took, and of the unexpected places it led me.

My quest was triggered by a series of impressive experiences of synchronicity in action. Many years ago I had published an astrology board game in Germany, which had led to me playing it on many occasions with many different people. What happened during these sessions was that the dice kept on turning up readings for the players that startled them and sometimes moved them deeply. What the dice was saying, it seemed, was mirrored in their current issues.

I recognised a potential here for using the synchronicity principle for self-development work. So, inspired by this idea, I went on to create an oracle book consisting of 1728 astrologically arranged sentences (12 x 12 x 12), read by throwing a twelve-sided dice. This led the reader through a series of sentences taken from different parts of the book which together created a narrative. I trusted it would reflect his or her current issues and lead to possible resolutions. And, while experimenting with the book myself, I found that the oracle principle worked – though not always!

The magic of synchronicity, it seems, only functions when we are tuned into the deeper levels of the mind. So, if the oracle book was being read in a desultory, or even worse in a critical way, the oracle would respond by throwing up meaningless sentences. And that's just what has prevented it until now from finding a publisher!

So it was synchronicity that gave me the first hints of how astrology might work, and it led me straight into the extensive opus of Carl Jung that I'd been avoiding. But it was in Jung's writings, while trying to grasp his model of the psyche, that I first met the archetypes which Jung defined as principles of pattern and meaning dwelling in the deepest ground of the collective psyche. A crucial point philosophically was whether these were just projections of the human mind, or really out there in the cosmos. And, as it turned out, in his later writings Jung spoke of them as if they were indeed cosmic factors inherent in existence as a whole.

Like everyone else I'd taken for granted that the outer world of consensus reality and my inner subjective life were two completely separate things – that is until I read about Jung's dialogue with his friend the physicist Wolfgang Pauli. At one point in their correspondence Jung compared the patterns that Pauli was exploring in the subatomic quantum field to the archetypes he was studying in the area of depth psychology. And Pauli commented to Jung that the flow of waves and particles on the quantum level appeared to him much more like the workings of a mind than of physical matter.[1] So what if both scientists were describing the same thing in different terms, and Jung's collective unconscious was identical with the quantum level in subatomic physics? The mind-boggling implication would be that what we know as reality is just a sequence of thoughts in a vast universal mind!

Pursuing this thread led me into postmodern science. Twentieth century scientific research, it seems, made a number of groundbreaking discoveries whose significance is only now beginning to filter through to the general public. Taken together they add up to a new scientific paradigm for the twenty-first century. And, what with the universe being grounded in the fluid quantum level, and concepts such as the nonlocality and the interconnectedness of all quantum particles implying a cosmic unity, it appeared to me that a new worldview was coming together in which astrology could have a place.

Above all Heisenberg's Uncertainty Principle, that revealed the necessary role of the observer in determining the nature of so-called objective phenomena, appeared to be relevant to my quest, as here was proof that we influence reality through the act of perceiving it. Its implication was that a subjective element must colour all our assumptions about what is 'out there'. The more I read about the mechanisms of human perception the more weird it all sounded, but the conclusion it led me to – that we play a role in creating what we take to be reality – appeared very relevant to astrology.

From all this I concluded that, in order to win a wider acceptance for astrology, it may not be necessary after all to change the shoe to fit the foot. We don't have to try to explain astrology through physical causes such as measurable magnetic influences from the stars and planets, because the foot (in the sense of how we see the universe) is now changing to fit the shoe. A new scientific paradigm is in the process of emerging that appears much more accommodating to astrology.

In the end it all boils down to the way we see the world, and that, of course, depends heavily on our conditioning. Most people in western society still take a worldview for granted that's based on the concepts of seventeenth-century mechanistic science. This is the so-called Newtonian scientific paradigm which rests on the law of material cause and effect, and takes linear time for granted as well as a fundamental separation between mind and matter. It replaced the earlier medieval paradigm that is found in the plays of Shakespeare. This earlier vision was a more organic world picture in which a 'chain of being' was the backbone of creation, and formed the basis for a pattern of cross-correspondences between different realms.

So are there other ancient cosmologies that offer an alternative to the scientific materialist worldview? What, for example, did our Neolithic ancestors believe about the cosmos? The archaeologists who measured the proportions and orientations of monuments such as Stonehenge have identified alignments in them to significant positions of the sun, moon and stars. I wondered if these could reveal something about pre-historic cosmology.

While visiting a number of megalithic sites in the British Isles in search of signs of Stone Age astrology, the idea came to me that what motivated these vast building projects could have been an attempt to mirror heaven on earth. In the light of this idea the Hermetic axiom of 'as above so below' took on a new significance. Perhaps, I decided, this was the main thread I needed to follow to unravel the mystery of how astrology works.

I visited the pyramids and temples of Egypt to research ancient Egyptian astrology, and like many before me I fell under the spell of that amazing civilisation. The wheels of time, established by eternal mathematical principles, turned predictably in the Egyptian sky. The goddess Nut, from whose womb the stars were born, swallowed the sun every night, but reliably gave birth to him again the next morning. And the divinity known as Maat, with her perfectly symmetrical wings, maintained an equilibrium in a cosmos that was adorned with order. How comforting it would be to shelter beneath her vast, soft wings, like a bird in a nest!

I became convinced that, contrary to scientific opinion, not only did the ancient Egyptians have an astrology from their earliest dynasties onwards, but their astrologers played a leading role in the social hierarchy. I also found I agreed with those researchers who argue that the advanced knowledge evidenced by the Egyptians in fields such as astronomy, mathematics and geodesy were the result of a legacy rather than an evolution, and were inherited from an earlier pre-historic super-civilisation.[2] Then could it have been the same people who were responsible for constructing all those Neolithic earthworks and henges across the length and breadth of the British Isles? In this case the key to how astrology works could be hidden amongst the numbers and proportions of the sacred geometry and astronomical alignments discovered in these structures.

I was now trying to get my head around things like whole and irrational numbers, ratios, Pythagorean triangles and – even more difficult to grasp – the relationship between number sequences and the musical scale. From Pythagoras and Plato, who I'd always venerated, I learned the difference between the quantitative and the qualitative dimensions of geometry, which brought me face to face with the problem of just how the inner and outer worlds are related. It also brought me closer to the answers I was seeking, because the figures of sacred geometry could be related to Jung's archetypes, and the geometric patterns created by the sun, moon and planets in the sky had a lot to do with how astrology works.

Then in 2006 the philosopher and astrologer Richard Tarnas published his major work *Cosmos and Psyche,* in which he creates a philosophical and psychological framework that brings together the two worlds of inner and outer reality.[3] And the pieces of the jigsaw puzzle I'd gathered so far began to slip into place. This was followed by Keiron Le Grice's *The Archetypal Cosmos,* which expanded on Tarnas' holistic cosmology, and put forward a conception of an archetypal cosmos as the new cosmological paradigm for the twenty-first century.[4] At last here was a cosmology that provided a context in which astrology could be justified!

It would not be clever if I were to reveal the solution I reached to the riddle of how astrology works in the Introduction. So if you want to find out about it you'll have to read the book. In Part One I follow a trail that leads me step by step towards my goal, whereas Part Two has a different focus. It explores the sources of astrological meaning in the context of the new paradigm of an archetypal cosmos.

Numbers and geometry, I'd discovered, form the basis of an ordering system of correspondences. And I now saw zodiac geometry as reflecting how the twelve astrological archetypes become the sources of astrological meaning through their mathematical relationships. The geometry of the zodiac, with its interesting parallels to the geometry identified in the ground plans of ancient temples, points us to patterns of significance that are cosmic in origin.

Therefore I explore how the meanings of the archetypes are expressed in a chart through its geometry and the symbolism of the chart components, and then go on to the practical application of all this. I describe a simple archetypal approach to interpreting charts, introducing the aid of archetypal fields of meaning expressed visually in diagrams. These support an approach to interpretation through the intuition and imagination, which can be a more rewarding way of investigating the meanings of the archetypes. Finally I give a demonstration of archetypal chart interpretation in action using the example of a famous person's chart. Have I captured your interest? Then read on ...

PART I

Archetypal Astrology and the New Cosmology

CHAPTER 1

A Paradigm Buster

The oracle principle

My first consistent experience of the phenomenon of synchronicity began in 1986 when I published an astrological oracle board game.[1] Over the years I played it with many different people, and was given ample opportunity to observe how the dice would confront them with readings that startled them. Apparently what came up for them during the game mirrored what was going on in their lives. This seemed to me more than chance coincidence. It was 'spooky action at a distance', to quote Einstein out of context, because there couldn't possibly be any cause-effect linkage between the numbers they rolled on the dice and the issues that were bothering them.

The famous psychiatrist Carl Jung had a similar experience back in the 1950s while he was experimenting with the *I Ching*. He noticed how the sticks, falling seemingly at random, always pointed to oracle texts that were relevant to his issues of the moment. And that set him thinking. Did another law exist capable of linking events beyond that of cause and effect?

The result was the publication of his monograph *Synchronicity: An Acausal Connecting Principle* in which he described the phenomenon of meaningful coincidence that he called 'synchronicity'.[2] He defined synchronicity as the coincidence in time of two or more causally unrelated events that share an overall pattern of meaning. One of these could be an inner and the other an outer phenomenon, or both events could take place in the outer world.

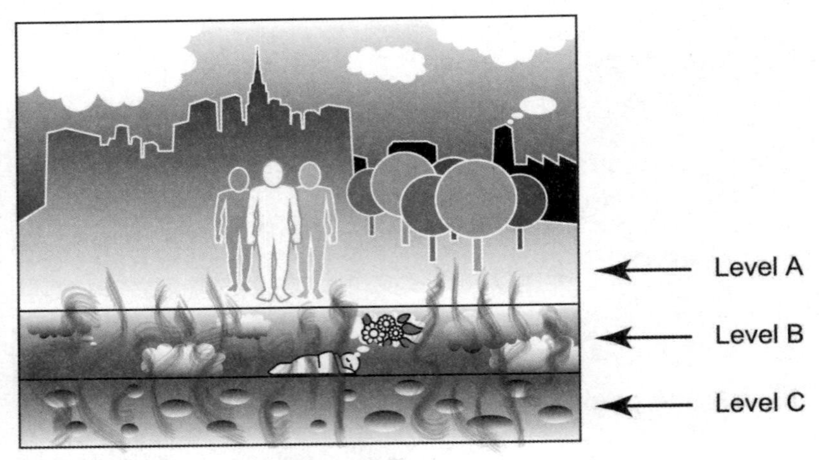

Level A: Matter, consensus reality
Level B: Mind, psychic reality
Level C: Quantum level, collective unconscious

Figure 1. Levels of reality.

Jung goes on to suggest that synchronous events with a similar meaning arise from the same source. Also, at times when we experience a whole series of synchronicities one after the other, we are witnessing larger and more universal patterns unfolding. If that's the case I wondered what their source could be and where could it lie? The thread I was following was now leading me into the world of physics, which I'd always found forbidding, but my wish to discover how astrology works was so strong I plunged in.

Figure 1 is my own graphic adaptation of the theories of the physicist David Bohm.[3] It's a model of reality showing my understanding of the relationship between what we experience as the two levels of the outer and inner worlds, together with the sub-atomic level of existence known as the unified quantum field. Level A, the outer world of things and other people, is where our attention is directed most of the time while we're awake. But we also live in the parallel world of Level B, composed of our inner life with its thoughts, feelings and imaginings. When we sleep Level A vanishes and we find ourselves wandering about on

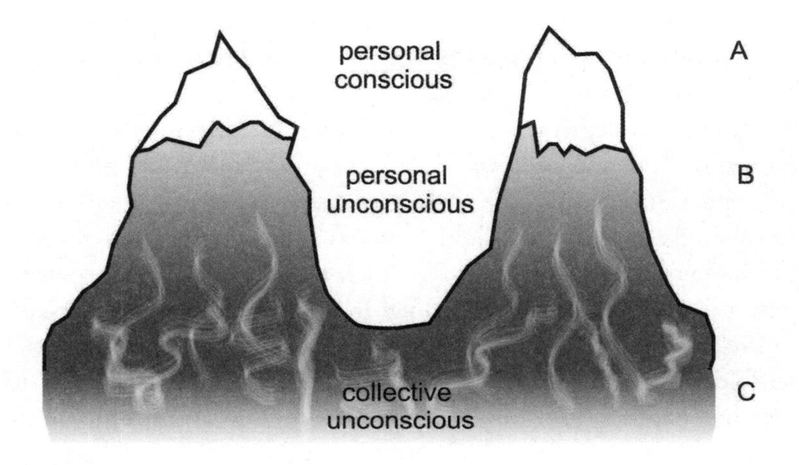

Figure 2. Jung's iceberg.

Level B in our dreams, unless, of course, we sink down to Level C in one of our deep sleep phases. Then we are at one with the common ground from which, as Bohm suggests, both Levels A and B emerge.

That Level B and Level C have been drawn below Level A in this diagram should be understood as a spatial metaphor. It accords with our experience of 'sinking' into sleep and 'rising' out of it again. But in fact the levels are not separate in this way, and Level C is not deeper than the other two in any spatial sense. It would be more true to imagine them as superimposed one upon the other.

In order to understand the individual mind in more depth (Level B), I looked into Jung's theories of the nature of mind. Jung, I found, believed the activities of the personal mind were determined by patterns within the deeper archetypal level of the collective unconscious. According to his model, the psyche consists of three levels – the personal conscious as the tip of the iceberg, below it the much larger area of the personal unconscious, and forming the ground of both the vast collective unconscious, which is an objective layer of mind accessed by the entire human race. The illustration in Figure 2 shows the minds of two different people according to this model, and how they are linked.

Jung's conception of the collective unconscious seems to correspond to the ground of existence that quantum physicists call the unified quantum field, represented in my diagram by Level C. Apparently this field contains what the physicist Werner Heisenberg called 'symmetries', which he understood as formative patterns of information embedded in it. And David Bohm speaks in the same vein of an 'implicate order' consisting of potentialities 'enfolded' in the ground of existence, which can then 'unfold' to become 'explicate' on Levels A and B. When in an unfolded state they become 'quanta', which are packets of coalescing particles. They are then sustained by the field for a while before collapsing back to become enfolded and implicate again.

So the levels in Figure 1 represent three dimensions of reality, which, although they are shown here as separate, in fact overlap. Symmetries unfold out of Level C to become explicate on either Levels B or A. On Level B, for example, they become ideas in the mind, and on Level A they can become natural objects such as a trees and animals. In the case of a house, the symmetries would unfold first as ideas in the architect's mind on Level B, and later, when they were expressed in bricks and mortar, become an object on Level A. Grasping these ideas moved me on in my understanding of the phenomenon of synchronicity. The main point was that the same pattern or idea could manifest on two different levels simultaneously – on the subjective level of mind, and on the objective level of material reality.

For example let's take a very common case of synchronicity: you happen to think of a friend you haven't heard from for months, and at that moment the phone rings and there she is on the line. This could be telepathy, but it could also be the unfolding of the subject of 'friend' on the two levels of an idea in your mind, and the outer event of a telephone call. Sometimes we experience a whole series of synchronous events. There are days when I keep getting small injuries by cutting, scratching or bruising myself, and then months can pass before I injure myself again. As an astrologer I would explain this by a transit of Mars – a synchronicity in itself – but it could also be seen as a pattern of mild self-mutilation unfolding from the implicate level.

Sometimes two or more events occur synchronously on Level A. For example family members may have the same birthday – a phenomenon that, as astrologers know, occurs in many more families than most people realise. The close coinciding of multiple birthdays within a family group is also much more common than statistics would predict, and is an indication of an astrological family pattern unfolding. This birthday synchronicity proves that dates of birth within a family group are not random, but demonstrate a wider pattern of significance in the process of unfolding.

Everyday magic

As an astrologer I encounter the magic of synchronicity every time I sit down to interpret a *chart* (see Glossary). In fact to be able to do this at all I have to trust the synchronicity principle. I rely on a correspondence between the configuration of the sky at the time of the birth (Level A, outer world), and the psychic patterns of the person born under that sky (Level B, inner world). And, because this is always the case – they actually do run parallel – I know from my daily experience that the ancient law of a correspondence between macrocosm and microcosm is still alive and kicking.

It's because of this mirroring effect that astrologers are able to look at the sky and use the level of visible astronomy to derive information about the psyche which is invisible but corresponds to it. Amazingly this works. What is more, as we become aware of synchronicities, we begin to notice that there are always correspondences going on between what we see happening out there in the world and our inner moods, thoughts and desires. The seventeenth century mystic Jakob Boehme knew this well when he said if you want to know yourself look at the world, and if you want to know the world look within.

Synchronicity is also at work linking the motions of the encircling planets to the unfolding of events on the world stage. Tarnas in *Cosmos and Psyche*[4] traces parallels across many centuries of human history between the cycles of the outer planets and synchronous world events. In my view he successfully proves that what happens on earth expresses astrological patterning. This kind of astrology is called *mundane* (see Glossary). It connects events that occurred separately and at different times, and puts them into wider contexts than that of material causation. In so doing it illuminates their significance within history's wider picture, and offers an invaluable overview of the larger patterns determining different time periods.

It was Jung's opinion that synchronicities only become meaningful if we have a context that allows us to see them within a greater order of meaning, and astrology provides this service. For example an interpretation of planetary *transits* (see Glossary) to our natal chart can reveal the meaning that current events have for us by putting them into the wider context of the phases of our life story. In doing so it gives us a bird's eye view from which we can survey our present, past and future, and thus come closer to an understanding of the fundamental dynamics of our life story.

Astrological interpretation requires the astrologer to focus her awareness before she can draw out the patterns of meaning in a chart. It is as if she must align her personal consciousness with the deeper ground of the universal consciousness for this to happen. And Jung saw synchronicity primarily as a phenomenon of consciousness, whereby a focusing and intensification of awareness was needed before the patterns could be recognised which, in a normal state of mind, we wouldn't notice.

While playing my astrological oracle game with different people over the years I'd discovered that the mood in which the dice is thrown is a crucial factor. Best was a relaxed and meditative state of mind in which normal awareness is heightened, and this is also borne out when we consult other oracles such as tarot cards or the *I Ching*. We need to be open to the feelings, thoughts and associations that are triggered by the oracle's statements, which are like pebbles thrown into a pool creating expanding rings of meaning. The oracle will not function properly if we are absent-minded or flippant or when we're critical and doubting.

I came to the conclusion that, when we are interpreting synchronicities in the context of astrology, the mind escapes from the limitations of time and space for a while. In the intensifying of our awareness, created by letting go and allowing our intuition free rein, we move down in our minds to a more profound level where the interdependence of things can be glimpsed.

To conclude this section: what I'd discovered so far was that astrology rests on the phenomenon of synchronicity. Synchronicity is part and parcel of how astrology works. Also it was now clear that the reason why astrology cannot be tested and proved by standard scientific methods is that synchronicity does not belong to the causal material level of the universe. It stems from a more inclusive level of reality in which matter and mind have their common source. Synchronicities also point to the cosmos being meaningful, as they would not occur in a random, purposeless universe. And neither would astrology work if a higher dimension as a source of meaning were absent.

Sir Isaac's secret life

Jung's theories of synchronicity had carried me forward in leaps and bounds, but I now needed to pause and take in the concepts I'd encountered. What is matter really and what is mind, and how are we to envisage a dimension in which they are unified? The consensus scientific picture of the universe in which I'd been raised had become destabilised by my experiences of synchronicity while practising astrology. I now needed a new world picture that could contain and explain them. So I started by examining the cracks in the Newtonian paradigm that I was in the process of leaving behind.

When the present materialistic, mechanistic account of a clockwork universe, referred to as the Newtonian model, came together in the course of the seventeenth century, the parameters it established banished astrology from science, and have marginalised it ever since. Newton is revered by the scientific establishment as the father of modern science, which is ironic because, privately, Sir Isaac was heavily into astrology. One famous anecdote tells of an occasion at the Royal Society when, under fire from a colleague for his belief in it, he retorted testily, "Sir, the difference between you and me is that I have studied the subject and you have not!" And there we have it. People who take the time to come to grips with astrology inevitably discover that 'there's something in it' – and much more than that!

Today the consequences of a number of twentieth century scientific discoveries, that long went unappreciated, are gnawing away at the roots of mechanistic materialism. Take for example the principle of cause and effect on which the whole Newtonian edifice rests. It decrees that every event must be preceded by a material cause. And Sir Isaac himself is said to have boasted that, if he'd had a chance on the first day of creation to ask God for the positions, masses and velocities of all the bodies, he would have been able to predict every subsequent event occurring in the entire universe. Of course he was talking about the visible universe, or Level A in Figure 1. Although there are many mainstream scientists around who believe that causal explanations will in time be able to explain everything that happens on Level B too.

In the eyes of these scientists astrology would only be acceptable if physical causes could be identified as its mechanism. According to their rules all findings must be based on facts proved by replicable empirical methods. And considerable research has been carried out by people eager to prove that the stars and planets *do* physically influence human life in ways acceptable to science. However the only measurable material influences of this kind so far established are those emanating from the sun and moon. Influences from the faraway planets and stars, which in astrology are seen as just as important, have never been properly accounted for in terms of physical cause and effect.

I wondered if Newton was privately into billiards too, because his laws of physics with their strict cause and effect framework apply very nicely to the billiard table. If a billiard ball is hit in the right way, and in consideration of all the relevant factors, it will go straight down the hole. However, as has been pointed out, this is an example of cause and effect on the most mechanical level of matter. Hit a man with a billiard cue, and you will not get such a predictable reaction!

Practising astrology is not a game of billiards; it's more like weather forecasting because, as is the case in meteorology, a very complex network of causation is involved. Weather situations, we are told, arise out of the flux of the whole, and therefore to forecast them synthetic rather than analytical thinking is needed. The same applies to astrology in that the truth of an interpretation has always to do with the relationships of all the multifarious factors to the whole. So chart elements must be understood in their widest context rather than through dissection and analysis. And that's why astrology cannot be adequately tested by the standard methods of conventional science.

Time gentlemen, please!

Then I had a revelation! I saw that the Newtonian rule stating that effects must follow causes depends on the existence of linear time. In a state of timelessness, there would be no past for the cause of a present effect to be in. And material causation is also dependent on time being objective. As Sir Isaac's later generation imagined things, on the first day of creation God set a gigantic clock in motion, which is still ticking today and will go on ticking till the end of the world. But meanwhile Einstein has come along and thrown a spanner in the clockworks by proving with his equations that time is neither objective nor absolute. Instead it's relative to the individual observer, who is only able to experience time within his own particular frame of reference.

Today most physicists doubt that time exists autonomously as something out there in the universe, and therefore separate from an observing mind. The physicist David Peat, for example, points out that we never perceive time as such. It's more like a concept we derive from our experience of change happening in the world around us. He suggests that, rather than being a single linear unfolding, time could consist of a whole spectrum of orders of which eternity and linear succession are just two aspects.[5]

The psychologist Roger Woolger compared the way we experience time as a procession of events to travelling through a landscape composed of different areas – woods, meadows and hills for example.[6] We register them one at a time as the road we are on winds through them, whereas, if we were in a plane overhead, we would be able to view the whole landscape at once spread out beneath us. When seen from a higher plane (pun intended), perhaps all our successive life experiences could be viewed as one totality. And, quantum physics assures us that this is indeed the case on the sub-atomic level, where concepts of past, present and future, and with them the cause-effect principle, break down into a state of timelessness.

I was interested to read that neuroscientists have proved our brain manipulates how we experience time in order to give us the experience of a one-way linear journey through life, with all its events arriving in a logical sequence. Now why should it do that? In any case their findings confirm that time is a psychological rather than a physical phenomenon, which adds another brick to my wall.

Is mind matter or is matter mind?

Twentieth century physics has disabused us of the idea that the atom is the elementary unit of matter, and, apparently, neither is the electron nor the quark nor any particle in particular. What everything boils down to at sub-atomic level is a chaotic soup of energy waves which only morph into particles when they are being observed.

This is the meat of Heisenberg's Uncertainty Principle, which proved that the act of observation by a scientist always affects what he's observing. Known as the 'observer effect', it effectively pulls the rug away from under the feet of all those scientists who claim their findings are objectively true. The observer must be included in the observed, the knower in the known. So, if Heisenberg is right, nothing can exist independently of an observing consciousness. And, if there were no observer observing, there would be no particles, no matter, no universe and therefore nothing whatever to see.

This all shows that mind and matter are much more closely associated than the commonsense view of reality leads us to believe. And secondly it implies that, in order to exist at all, the universe must require some kind of a conscious mind to be continuously aware of it, because without such an omnipresent observer everything would remain in a state of potentiality. Strange, because we all take for granted that things like houses, roads and trees remain in place when no one is actually looking at them – we rely on it being so. So is there a kind of super-consciousness out there keeping an eye on things, and making sure that all the waves continue to collapse into particles so that matter is sustained throughout the universe?

Pythagoras described a universal mind in which all individual minds partake, a mind that is present and active throughout the cosmos. So it's a very ancient idea. The Egyptians symbolised it with their image of the eye of Horus – a big brother continuously watching (see Figure 3). So is there is an all-embracing cosmic consciousness in which we all live and move and have our being. In which case have we found God? That's a question I'll leave unanswered.

Figure 3. The eye of Horus.

Bohm's 'implicate' order of potentials, that can unfold to become 'explicate', echoes Plato's theory of the 'Ideas' that exist in a timeless realm from whence they emerge to structure manifested reality. The rising and falling back of symmetries in Bohm's quantum flow, which looks chaotic at first sight, turns out to depend on patterns that are manifesting within the greater whole. Also, according to Bohm, what is destroyed is not lost but returns again to the enfolded, implicate state – which sounds to me like an account of the emerging and subsequent dispersing of ideas in a cosmic mind.

This makes sense to an astrologer, because the patterns she identifies in a mundane chart are seen in the context of the wider emerging patterns structuring history. Also patterns in a birth chart are understood as being potential before the child is born, to become actual at his first breath, and then to gradually unfold during the course of his life. Finally at his death they will enfold back into the flow.

Bohm was, I believe, the first to propagate the idea that the universe could be a three-dimensional hologram. In a hologram each component is a fractal of the whole and contains the whole encoded within it as information. This means that, enfolded within any particle of matter, there is information applying potentially to the whole universe, and there is also fractal correspondence on all levels of scale from great to small.

The idea of the nonlocality of particles, implicit in the model of the hologram, has repercussions for our understanding of how astrology works, as it suggests a basis for acausal connections. In the unified field containing waves with the potential to become particles, and symmetries with the potential to self-organise into forms, there is no 'place' as we know it, just as there is no 'time'.

A holographic universe could make sense of the ancient conception of the correspondence between the macrocosm and the microcosm on which astrology is based. According to the hologram model, patterns in human life as the microcosm would have fractal correspondence with the patterning of the positions

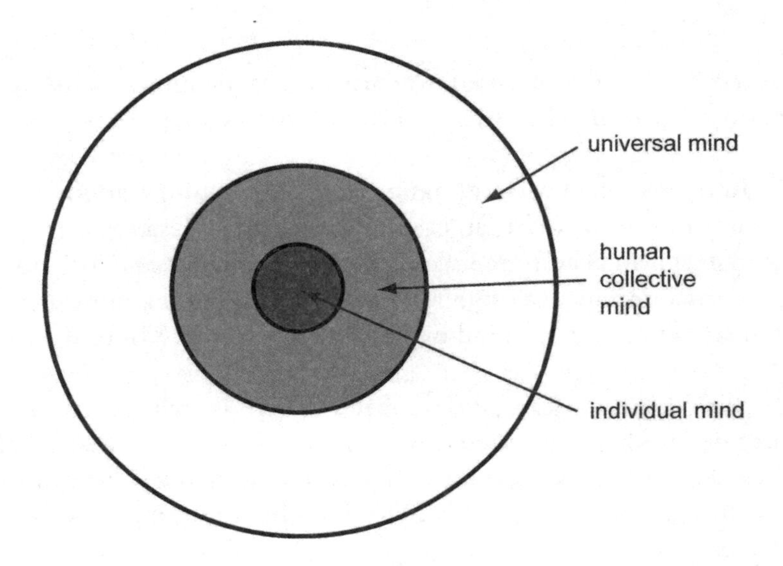

Figure 4. Levels of mind.

of sun, moon and stars in the sky as the macrocosm. So each individual person could then be seen as a mini-universe, and the universe as a whole would be like a maxi-person.

That takes care of matter, and next on the agenda is mind. Figure 4 is a diagram of mind represented as a nested hierarchy with its smallest unit – the individual mind – contained in the larger unit of the collective human mind, while both are embedded in the greater universal mind. The three levels should be imagined as superimposed upon one another as they interface, and thus no boundaries can be drawn between them.

If this conception corresponds to how things really are, then we can forget our pride in our intellectual achievements, because what we think of as *our* ideas may not be ours at all. As most creative people know, they could have floated into our mind from the minds of other people. Neither would we deserve credit for our deeper insights and revelations, which doubtless originate in the immense wisdom of the universal mind, a little of which may come filtering through to us. And human reason itself should be seen as merely a reflection of the rational faculty in creation. Within a fractally repetitive universe the human mind becomes the cosmic mind minimised, whereas the cosmic mind is a human mind full screen. The physicist Fritjof Capra wrote:

> In the stratified order of nature, individual human minds are embedded in the larger minds of social and ecological systems, and these are integrated into the planetary mental system – the mind of Gaia – which in turn must participate in some kind of universal or Cosmic Mind.[7]

And the American philosopher Walter Russell put it very succinctly when he stated that: 'all Mind is One Mind, that men do not have separate minds, and that all knowledge can be obtained from the Universal Source ... by becoming One with that Source.'[8]

We continuously experience as background noise what the philosopher William James called the 'stream of consciousness'. Voices in our heads go on babbling like a radio that's been left on. It babbles while we're awake and while we're asleep too, as we tune into it every time we dream. How much of its fleeting content we're aware of depends on whether our attention is focussed on it or not. Through thinking about an idea, we can actively extend it into a chain of thoughts, in which case we create what Heisenberg termed 'an observer effect'. What we are doing then is 'collapsing' waves in our stream of consciousness into concrete thoughts, and maintaining them through giving them our attention. But, when our attention is withdrawn, the thought forms collapse back into the inchoate stream flowing through our mind and become babble again – at least that's my experience!

The astrophysicist Sir Arthur Eddington once said that modern physics made the universe appear far more like a gigantic thought than a gigantic machine.[9] And, as we have seen, the German physicist Wolfgang Pauli in his correspondence with his friend Carl Jung said something similar. Jung on his side suggested that Pauli's symmetries, which Pauli had defined as loops of active information within the quantum field from which the forms of material things unfolded, were of the same order as the archetypes he had identified in the psyche, and which he saw as formative principles within the collective unconscious. He wrote:

> Our psyche is set up in accord with the structure of the universe, and what happens in the macrocosm likewise happens in the infinitesimal and most subjective reaches of the psyche.[10]

And that is just what astrologers have always been saying!

CHAPTER 2

Cosmologies Ancient and Modern

Stone Age cosmology

The Greek word *kosmos* means 'beautiful order'. Thus, when we use it to describe the totality, an integrated whole is implied that has quality and value. On the other hand 'universe', a word of Latin origin, means a unity subject to law, and using it suggests that the totality is subject to laws that are knowable.[1] The former approach is intuitive; the latter more rational and scientific. Both terms, however, presume the presence of organising principles within the whole, and cosmology is the study of these principles. For purposes of style I will use the two terms interchangeably in what follows.

Our conscious and unconscious assumptions will always influence what we see and how we interpret it. For example the materialist neo-Darwinian conception of Nature has shaped our understanding of ourselves and our planet, while in other cultures the world is seen differently. Our ancestors who inhabited the earth thousands of years ago also must have looked at it through very different eyes.

We take our conception of reality for granted, however the findings of modern science suggest that a large part of what we believe to be real is projection. We 'project' what we expect to see – though this is not to say that there's nothing out there. Our impressions are likely to be a composite of what we are expecting to find and what actually exists.

Those scientists of a materialist bent, who believe that every natural phenomenon is explainable in terms of matter and natural forces, remain confident that all the laws of nature will one day be known. However they are presuming we live in a universe with fixed functions, and this is not the universe that Einstein describes. His theory of relativity together with the findings of quantum physics point to the universe being an ongoing process rather than a static object. So the nature of reality as we perceive it may be constantly evolving, and we may literally live in a different cosmos than that of our grandparents. As Le Grice writes:'Humanity's spiritual odyssey is taking place within a universe that is itself creatively evolving and continually complexifying into different forms.'[2]

As I see it astrology has come down to us in a fragmented form from a civilisation that lived on this planet in remote pre-history, whose knowledge of maths, astronomy and metrology was amazingly advanced. John Anthony West in his book *The Case for Astrology* agrees. He writes:

> Traditions but no historians maintain that astrology was one of the products of 'divine inspiration' of the 'Sages of Atlantis'. This knowledge, it is claimed, in one form or another passed to Egypt, China, India and the vanished civilisations of South America.[3]

We have been taught to take a linear, evolutionary version of history for granted, and many believe the people who lived before the Greeks were little more than cavemen. Human progress is peaking now, as proved by the scientific and technological achievements of the present age – at least so we've been led to believe, though I don't buy into the myth of progress. While acknowledging the value of modern scientific and technological know-how, I still think the people who built Stonehenge and the Great Pyramid were wiser than us in many ways, and also that human civilisation is thousands of years older than the cultures of Greece or Rome.

The fact that ancient myths and sacred histories the world

over conform to similar patterns is taken by some as evidence that an advanced civilisation of worldwide reach originally lived on this planet. Also it has been shown that the oldest temples and monuments so far investigated were designed in accordance with one uniform scheme of proportion, and in units of measurement that are the same the world over.[4]

Research carried out into the numbers inherent in the ground plans of sites such as Stonehenge and Avebury has proved that the Neolithic mathematicians of the British Isles used the same system of maths as their ancient Egyptian and Babylonian counterparts. This all could point to these cultures having common roots in an ancestor civilisation now vanished in the mists of pre-history whose knowledge they inherited.[5]

Plato is reported to have said, 'the ancients were wiser than us'. And, as he was born in the fifth century BC, this begs the question of which ancients he was referring to. When the Greek historian Herodotus visited Egypt around two hundred years before Plato, he was told by a temple guide that Egyptian history went back thousands of years, and that the sun had twice risen where it now sets, and twice set where it now rises during that time.[6] An astrologer would make that one and a half precessional cycles amounting to 30,000 years in all![7]

However no archaeological remains have been reported from the Nile valley to suggest the presence there of a civilisation that far back. But suppose settlers from another part of the globe moved to this area, attracted by the Nile valley's fertility. And suppose this happened after the legendary land of Atlantis sank in the ocean waves during a natural catastrophe, which some writers believe occurred around 10,000 BC. All this of course is conjecture.

The cosmology and metaphysics that inspired these ancient civilisations now appeared relevant to my quest into how astrology works. Perhaps the ancient ones in their time understood more about the cosmic laws on which the truth of astrology rests than we can glimpse from inside the shell of our matter-based cosmology. So was there astrology in the Stone Age? That had become my leading question and, following the clues they left behind, I tried to piece together an understanding of the cosmos as seen by the builders of the megaliths.

First there are the astronomical alignments to sun, moon and significant stars planned into the monuments erected on sacred sites in the British Isles between 4000 and 2500 BC. Together with the numbers and geometry inherent in their designs, different researchers have shown how these can be used as keys to unlock the mysteries of what looks very like a prehistoric astral religion. For example the engineer Alexander Thom, who back in the 1950s was the first to carry out a thorough investigation of the ground plan of Stonehenge, discovered geometric figures within it that are based on Pythagorean triangles. These triangles include a right-angle and have sides that are respectively 3, 4 and 5 units long – a geometric figure considered by the ancients to be especially sacred.

The alternative Egyptologist Schwaller de Lubicz, who pioneered the symbolist approach to the culture of ancient Egypt, was pursuing a rigorous investigation into the designs of the major Egyptian temples while Thom was investigating Stonehenge. He discovered a canon of proportion in them based on the geometry of the human body and on the intervals of musical harmony.[8]

Also, according to the researcher Paul Devereux, the very stones from which megalithic monuments and temples were built have acoustic properties, which make significant the way they are arranged.[9] As both musical harmonies and geometric shapes are reducible to numbers, and numbers are a universal language intelligible to all intelligent creatures, is has been suggested that this knowledge was encoded in these structures to communicate with future generations – with us for example!

The Victorian astronomer Norman Lockyer was the first to investigate the astronomical alignments of Egyptian temples. He discovered, for example, that the rays of the rising sun at the summer solstice entered the gates of the temple of Amon-Ra at Karnak, and shone down its central axis to illuminate the innermost sanctuary. The earliest alignments known in Egypt are built into the Nabta stone circle in upper Egypt, which was constructed at the latest around 4000 BC, and which, it has been suggested, may not have been built by Egyptians at all, but by the civilisation responsible for the look-alike stone circles of Atlantic Europe.[10]

The Newgrange passage-tomb in Ireland presents the earliest example of a solar alignment known of in the British Isles. Its entrance is aligned with the sunrise point at the winter solstice. On this day, and for a few days before and afterwards, the sun's rays pass down the central passageway to illuminate a stone in the inner chamber carved with a design of mystic spirals. On all the other days of the year this inner chamber is in total darkness.[11]

Such astronomical alignments certainly had an astrological purpose – in the sense that they had symbolic meaning – and must have figured in the religious beliefs of those who built them. Perhaps the following quotation from the Pyramid Texts of Unas is a clue to what was going on. These sets of hieroglyphics, which date back to around 3000 BC, are the earliest examples of sacred writing as yet found in Egypt.

> I have walked on thy rays as if on a stair of light to ascend to the presence of Ra. Heaven has made the rays of the sun solid so that I can elevate myself up to the eyes of Ra ... they have built a staircase leading to the sky.
>
> ... I have trodden those thy rays as a ramp under my feet whereon I mount up to that my mother, the living Uraeus on the brow of Ra.[12]

So could the alignments have been made to create 'pathways' linking places on earth with features in the sky? If so, the alignment from the shaft in the king's chamber in the Great Pyramid to Orion's belt could have been intended as a route for the spirit of the deceased Pharaoh to take on his post-mortem journey to the stars. Alignments like these, it has been suggested, could have also been used by living shamans space-travelling in their astral bodies.

And if they also functioned as corridors by which the astral powers could regularly descend to earth, the traffic along these pathways would have been two-way.

In the nineteenth century, astronomer Norman Lockyer suggested that the alignments he discovered in many Egyptian temples towards the sunrise positions at the solstices or equinoxes were created to bind the light of the sun at times when it was especially 'potent'. Alignments were also created to specific stars so their light could be 'captured' at regular intervals as they passed overhead.[13]

According to John Michell, ancient temples served both as accumulators of cosmic forces and as instruments for transmitting these forces into the environment. In other words they were like 'central power stations' for their areas. And the currents of harmony they radiated were maintained 24/7 through the ceaseless activities of the priests, astrologers and officials. It was some commitment because the prevalent belief was that, if the harmony should cease to be preserved, defects could appear in the state that would undermine it.[14]

Legend tells that communion with the stars was the highest form of meditation practised by the Pythagorean brotherhood. And in this vein a later Gnostic writer described how the stars and planets serve to nourish the earth and its inhabitants, waxing lyrical about:

> The sumptuous majesty of the night, adorned with a clear light, though one less than that of the sun; while the other mysteries moved severally amidst the heavens according to fixed movements and periods of time, ordering and nourishing the lower realm through secret effluences.[15]

I came to the conclusion that these ancient cultures envisaged the cosmos as a giant organism alive with gods. It was a living whole in which all things were interconnected as in a web. And this essential unity came about because, ultimately, everything in creation was the expression of one divine mind. Thus, in the eyes of the Egyptians, the world with all its objects, its people, its events, its visible and invisible dimensions was not only empowered by, but also contained in the vast consciousness of the creator god Atum whose name means 'the totality'.

It is likely that the ancient Britons held a similar holistic view of creation. While the pyramids and temples we know today were being constructed in Egypt, they were erecting their monuments of earth, wood and stone across the length of the British Isles. The human collective unconscious knows no geographical boundaries. And, in place of more tangible forms of information exchange, telepathy via the worldwide-web of the collective unconscious could be the explanation for the phenomenon of people in different parts of the globe synchronously adopting similar views and customs.

'All things are full of gods,' the Greek philosopher Thales is reported to have pronounced, expressing how the ancients experienced Nature, as replete with divine powers both in its animate and inanimate forms. In Egypt natural processes were believed to be fuelled by the ubiquitous life force represented by the god Osiris, who caused the corn to grow after the annual flooding of the Nile, and brought fertility to the land. Humankind together with all living creatures shared in this force, just as they participated in all the divine cosmic powers, because for the Egyptians, there was no essential separation between the divine, the animal and the human.

The ancient Britons may have had their Osiris too and called him the Green Man, familiar to us from Celtic myth. Like Osiris he also passed through the perpetual cycle of birth, fruition, death and rebirth on which life hinges. Thus the ceremonies at the seasonal festivals held at Stonehenge and Avebury could have been orchestrated to take their participants through the symbolic process of his death and resurrection, as did the similar Egyptian festivals honouring Osiris, for the purpose of renewing and revitalising the community and natural environment.

In ancient times people experienced the universe as 'ensouled'. Living in a symbiosis with nature, they were able to palpably feel the presence of the spirits in the trees, the rocks, the rivers and the mountains surrounding them. Today we live in a disenchanted world. To modern scientific atheists, immune to symbolic thought, the universe is a soulless, meaningless machine. Unless they've questioned the scientific materialistic paradigm and gone beyond it, how can Egyptologists grasp the spirituality that inspired Egyptian art and architecture, or begin to understand the true functions of their temples and pyramids?[16]

Spirit was seen as primary in the ancient cultures, and matter, which derived from it was secondary, whereas today it's the other way round. Science proclaims that the laws of matter determine what we are. We've been taught that we were formed by our genes, and that our psychology is the result of our brain chemistry – after all drugs have the power to alter our moods. Therefore for our generation it follows logically that consciousness is an 'epiphenomenon', a mere by-product of the activities of the physical brain, whereas in ancient times consciousness was seen as the very stuff of which the universe is made.

The cosmos was alive for the ancient Egyptians, and not only in the sense of being replete with spiritual powers; it was an active, intelligent being in its own right. Nicholas Campion, writing in *The Dawn of Astrology* puts it this way:

> The Egyptian cosmos was alive; it was at once sexual and spiritual, physical and religious, enchanted, magical and scientific ... and, at its core, the Egyptian cosmos was moral.[17]

That the cosmic being also possessed an aesthetic sense was evident from the inherent beauty and harmony of plants, animals and all natural forms that Egyptian artists and craftsmen emulated so reverently in their art and architecture.

The ancient canon of proportion

The reason cosmology was considered the chief of all areas of study by the ancients was because the understanding of the cosmic order it offered provided the model for a harmonious society. John Michell writes, 'the most cherished possession of every race was its sacred canon of cosmology, embodied in the native laws, customs, legends, symbols and architecture as well as the ritual of everyday life.'

And he goes on to demonstrate the importance of the concept of the temple, which he understands as a model for what he calls 'the living canon of proportion', 'for the temple was itself a canonical work, a model of the national cosmology and thus of the social and psychic structure of the people'.[18]

As late as the fourth century BC, Plato was witness to the fact that the Egyptians still maintained the ancient canon, and wrote in *The Laws* that, as long as it's adhered to, it ensures the creation and maintenance of divine order on earth. Michell claims that the entire philosophy and science of the ancient world was founded on the principle that the cosmic ideals of harmony and proportion were to be emulated in human life. These were clearly demonstrated in the order of the sky by the moving geometry of sun, moon and planets. Therefore the Egyptians, and very likely the ancient Britons too, mirrored this conception of universal order in their architecture, and strove to maintain it through their calendars of feast-days and ceremonial ritual.

In Egyptian theology it was the goddess Maat who stood for this order. She was so finely balanced that the feather on her head always remained motionlessly upright! And it was her responsibility to maintain the gold standard of justice and proportion, by prescribing just bounds that were both physical and moral. Maat would ensure that, after phases of social conflict bringing chaos, order returned once more to the land of Egypt. And she also had a presence in the Judgment Hall, where the hearts of the deceased were weighed on the scales against her feather of truth. Only the extremely light-hearted survived this test; those with heavy hearts were thrown to the waiting crocodile!

The temple's calendar of feast days, each with its prescribed agenda of ceremonies, had been astronomically calculated in even more ancient times, according to the cycles of the heavenly bodies. However, the sun, moon and planets, although each moves with reliable regularity, interact differently at different times within their cycles. And this produces new situations in the sky for the astrologers to interpret, which explains why they enjoyed such high status in the Egyptian temple hierarchies. Only they possessed the necessary knowledge of celestial dynamics, and so could ensure that the pattern of temple ritual was correctly adapted to each new astronomical situation.

As reported by the early Christian theologian Clement of Alexandria (150–215 AD), on Egyptian feast-days the chief astrologer would walk at the head of ceremonial processions immediately behind the Singer who would lead them:

> The Singer is followed by the Horoscopus bearing in his hand the measure of time and the palm branch – symbols of astrology. It is his duty to be versed in and recite the four books of Hermes which treat of that science. Of these, one describes the positions of the fixed stars, another the conjunctions and illuminations of the sun and the moon and the others their risings.[19]

The architects and surveyors, who drew up the ground plans for the Neolithic temples based on the ancient canon of proportion, must have enjoyed as much kudos in society as the astrologers. Not only were the geometric and numerical factors they dealt with of vital significance, but they were responsible for the temple's astronomical orientation, as well as its accurate positioning in the surrounding landscape. Michell also claims that temple sites were chosen with reference to terrestrial magnetism and earth currents, which the ancient dowsers were proficient at measuring. And when a group of monuments were laid out in what has become known as a sacred landscape, such as the Brú na Bóinne in Ireland, or the area around the Ring of Brodgar in Orkney, alignments between the different structures in the complex were accurately calculated over large areas.

Each temple was laid out in relation to the four cardinal directions as well as to significant points on the horizon such as the rising and setting positions of the sun, moon or certain bright stars. Great care was taken in this task because it was believed that, only if the alignments were accurate, would the energies bearing on the temple further its sacred purpose. For example, an alignment in Egypt to the heliacal rising position of Sirius, on an energetically loaded day of the year such as the summer solstice, would enable Isis, the deity of this star, to descend and 'appear' in the temple's halls and sanctuaries. It should be noted, however, that in themselves the stars were not seen as gods by the Egyptians; rather they were the dimensions of the gods that are visible to human eyes.

Thistles and red peppers

Among the Hermetic books, which were based on manuscripts taken from the Egyptian temple libraries, was said to be a work that catalogued all things under heaven according to a set of ruling principles.[20] This was essential reading for Egyptian physicians, whose medicine was based on the energetic attractions or 'sympathies' between objects from the different realms of nature. For example thistles, iron, the ram, the planet Mars, red flowers and red peppers would all have been grouped under the same principle that astrologers today call 'Aries', and whose correspondence in the human body is with the head.

The doctrine of the correspondences, which derives from the ancients' conception of the cosmos as an interconnected whole, links items from the three realms of sky, earth and human life together according to the law of like attracts like. 'All phenomena in the divine and material realms alike are linked together by "sympathetic" powers or energies into one pleroma', is how Garth Fowden puts it in his book *The Egyptian Hermes.*[21]

The term used for this ancient system by which all things are classified is the 'great chain of being.' It was a conception that derived from the pre-historic canon of knowledge, and survived in alchemy and medicine into the seventeenth century of our era, while it still remains central today to astrological theory. In fact astrology can be seen as the ark in which this portion of ancient wisdom has been preserved over the last four hundred years from the flood of scientific materialism.

The Egyptians assigned the different categories of phenomena to fundamental archetypal powers they called 'Neters' *(netjer* in Egyptian) – although the word 'assigned' is not appropriate here, as the allocations were defined by cosmic law. The Neters are pure, abstract principles that dwell on a higher rung than the gods in the Egyptian hierarchy of divinities (see pp.78–9). Correspondences arose between things governed by the same Neter, which attracted and held them, as it were, in its field of influence. In this respect the Neters are very similar to, or are identical with, the astrological archetypes as we are going to see.

These correspondences played an important role in magic, where they were used to summon up daemons and nature spirits and coerce them into carrying out tasks. Shakespeare echoes this practice in his play *The Tempest* where the magician Prospero binds the nature spirit Ariel and forces him to do his bidding. Magic works in my view because, on a deeper cosmic level, the human will is linked with the all-powerful cosmic will. Therefore a human intention can be boosted by the greater power of the whole, if the secret of how to engineer this is known by the person holding the intention.

The use of astral magic in association with psychic powers, whereby a magus would seek to engage the stars and planets for his purposes, was widespread in Egypt. Many centuries later at the time of the Reformation, the theologian Cornelius Agrippa described how, through an intense focussing of the mind, human consciousness can be brought to merge with the consciousness of the stars, and even with the consciousness of the intelligences that rule them. The sage, the magician and the astrologer, it seems, had their techniques for appropriating the powers of the cosmic forces by consciously uniting with them, and, by aligning their wills with these greater cosmic wills, became participators in the creativity of the cosmos.

The love affair between earth and sky

As space exploration using ever more powerful telescopes reveals our planet as a mere pinpoint in a universe that is composed of trillions of other planets, and suns, and galaxies, the mind boggles at the vastness of it all, while inside we feel meaningless and empty. How differently the cosmos appeared to our Stone Age ancestors! The mythologist Joseph Campbell (1904–1987) described this beautifully in the following passage:

> The mystery of the night sky, those enigmatic passages of slowly but steadily moving lights among the fixed stars, had delivered the revelation, when charted mathematically, of a cosmic order … A vast concept took form of the universe as a living being in the likeness of a great mother, within whose womb all the worlds, both of life and death, had their existence.[22]

The ancient Egyptians lived beneath a sky imagined as a nurturing mother. She was the graceful goddess Nut, whose body was studied with the stars of the Milky Way.[23] The most important correspondence in the ancient system was that between sky and earth, portrayed as a love affair between Nut and Geb (see Figure 5). In those days Nut, the sky, was female, and Geb, the earth, was male. Today we automatically imagine a male sky and a female earth, which just goes to prove that there's flexibility in how even the most basic archetypes manifest, a subject we'll be returning to later.

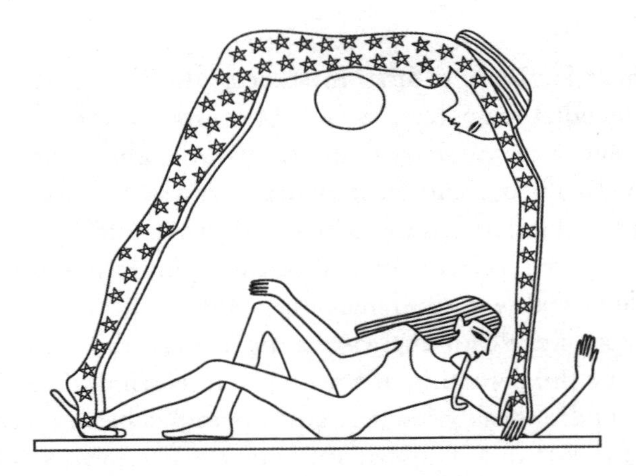

Figure 5. The lovers Nut and Geb.

When Nut and Geb copulated they created a channel for the life force to flow down from the sky to fertilise and vivify Nature. The Egyptians envisaged this force as ubiquitous – an energy that was everywhere, even filling the spaces between the stars – and the planets played a role in disseminating it.

Nut and Geb, who in the beginning were one flesh, had been torn apart which must have been a traumatic separation for them, because ever since then according to the myth they've been yearning to re-unite. The loss of an original oneness between heaven and earth must be a very ancient conception. Possibly this is what motivated the ancient Britons during the fourth and third millennia BC to undertake the construction on such a massive scale of earthworks and henges across the length and breadth of the land.

Our Neolithic ancestors may have invested so much time, energy and resources in these projects because they believed they were healing the rift between heaven and earth, and therefore between the gods and men, through creating structures on the ground that mirrored the sacred features they saw in the sky.

The Greek historian Herodotus reported that Egypt was first ruled by the gods; then, after many millennia, by 'the companions of Horus', and only several thousand years after that by the

Pharaohs.[24] Similarly in Irish mythology the Tuatha dé Danann, who descended from the sky, were the seen as the original race of gods and were referred to as 'Lords of Light'. These legends accord with Plato's statement that the gods once lived on earth with men and instructed them in government. And it is these gods who are reputed to have established the canon of temple law as a complete instrument of reference.

But then, he tells us, there was a rupture and the gods departed. Perhaps it's the departure of the gods that is referred to when the ancient myths speak of a separation between heaven and earth. And all those efforts that were made by the Neolithic populations to align the earth with the sky through their immense building projects may have been motivated by the hope that, if heaven could be created on earth, the gods would be seduced to return.

The big secret

Before the sky could be mirrored on the earth it had to be known and its workings understood. And this need furthered the development of astronomy to study the body of the universe together with astrology to study its soul. I concluded from my Egyptian studies that astrology played a central role in Egyptian civilisation from the earliest dynasties onwards, although many historians and orthodox Egyptologists would not agree with me there. However four great astrological works were among the collection of sacred literature that was preserved over the millennia in the main Egyptian temples, and these are the books that Clement of Alexandria described the Horoscopus as carrying in the feast-day processions.

For millennia access to these books had been strictly limited to the priesthood. As the Greek physician Hippocrates (460–370 BC) wrote: 'But holy things are shown to holy men. The profane may not be shown them until they have been initiated into the rites of Science'.[25] However, when the Greeks took over after Alexander the Great conquered Egypt in the fourth century BC, the temple libraries were raided and their contents removed to stock the new library the Greeks were building in Alexandria. Thus the long preserved and carefully guarded contents of the Egyptian sacred books entered the public domain.

In the centuries that followed various Greek mathematicians and scientists took the credit for discoveries they'd no doubt plagiarised from Egyptian temple library manuscripts. One of these manuscripts could have been the *Emerald Tablet of Hermes Trismegistus,* which, although it emerged much later when Egypt was under Arab rule, betrays its ancient Egyptian origin in its style and content.[26] The following quotation from the opening sentences struck me as being very relevant to how astrology works:

> 'Tis true without lying, certain & most true. That which is below is like that which is above & that which is above is like that which is below to do the miracles of one only thing. And as all things have been and arose from one by the mediation of one: so all things have their birth from this one thing by adaptation.[27]

Thus the tablet drops a bombshell in its first line by revealing the secret of the magical correspondence between the patterns of the heavenly bodies in the sky and the patterns arising on earth – that which is below corresponds to that which is above and vice versa. The microcosm is mirrored by the macrocosm. Its opening sentence puts a very strong emphasis on the statement that follows as if anticipating that practically minded people, then as today, are not going to believe it. Incredible as it sounds, it goes on, in truth and without any deceit, what follows is certain and correct ...

For thousands of years the big secret had been guarded by temple adepts, all too aware of the power it bestowed and how this power could be abused, but now it was out. Thus astrology passed from the reverend hands of the Egyptian temple priests into the market place, where, instead of being employed for sacred purposes, it was used for divination for political gain and for vulgar fortune-telling.

In a nutshell the big secret contained in the *Emerald Tablet* revealed that the mutual mirroring of earth and sky should be understood quite literally. The Babylonian priests were also in on this secret, which led to some centuries of precise and consistent record-keeping by their astrologers of the correlations between the motions of the heavenly bodies and events in the human world. The results of their work lie buried in the British Museum vaults, which house a collection of hundreds of clay tablets from the library of Ashurbanipal, an Assyrian king from the seventh century BC.

However, apart from anecdotal evidence, we have no concrete proof that astrology was practised in Egypt at this time or earlier. It would nevertheless be in accordance with the careful and methodical way affairs were conducted in the Egyptian temples for the astrologers to have kept records. Herodotus in his *Histories* asserts that the Egyptians 'have always kept a careful written record of the passage of time', and then again that 'the Egyptians ... by their practice of keeping records of the past have made themselves much the most learned of any nation,'[28] which point to this being the case.

However, as I see it, the truth of 'as above, so below' was not discovered piecemeal through the accumulated results of thousands of nightly observations of the sky. This conventional account of how astrology developed comes from a mindset influenced by the 'bottom-up' procedures of analytical science that start with discrete facts and work up to general truths. I prefer a Platonic 'top-down' explanation whereby knowledge of the big secret and how to make use of it was part of a legacy. A body of knowledge was inherited by the Egyptians and Chaldeans from adepts whose lineage went back to that lost earlier civilisation. These could have been those very sages referred to as 'the companions of Horus' in Egypt and the 'Tuatha dé Danann' in Ireland.

A better cosmic fit

Our view of the universe is changing. A new paradigm is coming in to challenge the mechanistic, scientific worldview. Slowly but surely the cosmos as seen by the ancients is beginning to re-emerge. Processes that science has been interpreting as the rule-bound functioning of mechanical cause and effect, such as how our genes influence our behaviour, are now allowed more flexibility (as for example in the new field of epigenetics). And quantum physics has shown that the material level of creation, that has been the main focus of science for centuries, is not as ultimate as people once believed. It turns out to be only the surface of an immense ground extending down to levels where matter and consciousness merge.

The present trend towards a more holistic cosmology was can be traced back to the 1970s when the 'Gaia hypothesis' was advanced by the scientist James Lovelock. It led to the personification of our planet as the primordial Greek earth goddess Gaia. Gaia theory is holistic in that it sees all aspects of the planet as components of one overall system – a view of things echoed by the new science of deep ecology. A holistic, organic worldview inspires the ecological movement in its present campaign to support Gaia by counteracting the effects of species decimation as well as of global warming and climate change.

Quantum theory has caused a revolution in cosmology by presenting the universe at its most fundamental level as an unbroken whole in which everything is interconnected. This account of the seamlessness of the sub-atomic level of existence is undermining the belief in the dualistic separation of mind and matter that was hammered into the collective unconscious by Descartes and other seventeenth century scientists and philosophers.

A new perspective on the cosmos has also been contributed by the findings of systems science since its development in the 1980s. The earth is seen by the scientists in this field as a nested hierarchy of systems, with lesser systems contained in and determined by greater. And their all-inclusive model is used to describe our relationship with the solar system, the galaxy and the universe as a whole.

Bell's theorem of nonlocality also accords with these views in proposing that every event that occurs results from the influences of the universe as a whole, and can also influence every point in the universe simultaneously. So instantaneous connections occur at wave and particle level, and distances in space and time then become irrelevant. Heisenberg's Uncertainty Principle has also led to the questioning of objectivity as such. If an observer inevitably influences an event simply by the act of observing it, then in a sense he is its co-author. And this observation becomes one of the fingers that point to a co-creative role for humans within the cosmos.

So it seems that the way the ancient people saw the cosmos, as a living, self-determining organism, could be making a comeback. The astrologer Dennis Elwell noticed this trend back in the 1980s when he stated in his prescient book *Cosmic Loom:* 'The unified universe, its parts all holographically reflecting the whole, must itself be inherently conscious.'[29] And Richard Tarnas in *Cosmos and Psyche* describes the cosmos as one great living being, 'throbbing' with vitality and consciousness. He goes so far as to propose that the universe not only has a body, but also possesses a mind and a psyche – a view that recalls the ancient conception of an *anima mundi.*

Astrology appears as superstition when judged according to the parameters of scientific reductionism. There's no place for it in that worldview as it is impossible to test and prove using conventional scientific methods. That's why it's been outlawed since the Newtonian paradigm took hold of the collective mind. However, as we have seen, the worldview on which materialism is based is now being undermined by postmodern science, and a new vision of the cosmos is emerging within which astrology could be reinstated.

It fits quite happily into a cosmic view that recognises the interrelatedness of all things, as this provides a philosophical basis for the astrological correlations. And, if all the cosmic realms are essentially interconnected, the ancient theory of correspondences so basic to astrology finds its justification. Astrology thrives among people who believe that consciousness rather than matter is the stuff of reality, and the idea that individual minds all share in the one great universal mind, partaking in its thoughts and intentions, would explain how collective astrological patterns are mirrored in individual lives.

All these understandings had moved me forward towards my goal of explaining how astrology works. Above all, if the universe is envisaged as possessing a rational intelligence and a feeling soul, a context of meaning is provided for the events occurring in human life, and this is what astrology is all about.

The universe as a living creature would have the freedom to be creative and also to change its mind, and so we must reckon with a degree of unpredictability. We can imagine it having a durable central structure, like the trunk and main branches of a tree, whereby the leaves and blossoms that emerge from its twigs will all have different textures and change with the seasons. The Neo-Platonist philosopher Plotinus found an even better metaphor. He compared the ordering principles of the cosmos to the patterns of a dance and the universe as a whole to a dancer.[30]

I had come to two main conclusions at this stage in my quest – that astrology is concerned with the qualitative dimension of the principles that structure the cosmos; and that the big secret – the law of 'as above so below' – rather than being a poetic metaphor is a literal truth. In fact it might be the most fundamental cosmic law of all!

CHAPTER 3

Divine Numbers

Numbers sacred to the gods

In the ancient world it was believed that a mathematical code underpins everything in creation, and that all the forms of Nature derive from it. Its mathematical sequences lie behind the geometric shapes that repeat on all levels of scale from the subatomic to the galactic. In other words the same hexagram that structures the tiny snowflake also manifests in the hexagram discovered in the vast gas clouds swirling round Saturn's North Pole.

According to the astrophysicists these six straight lines of more or less equal length, which are clearly to be seen in NASA photographs, were created by random gas molecules self-organising. But this kind of explanation would not have satisfied Plato. He'd have seen it as the material manifestation of an eternal geometric form.

We experience the code's number sequences audibly when we listen to music, which then become sounds we can hear. Pythagoras taught that the musical harmony to which our organs of sense have access is an expression of cosmic harmony. Similarly, when they are expressed as geometry, numbers become shapes that we can see. And, once they've become perceptible to our physical senses, they acquire emotional and psychological significance, and then have quality as well as quantity.

The qualitative dimension of numbers has been completely lost sight of today. However, when arithmetic and geometry were taught in the academies of Egypt and Greece, the study of the qualitative dimension of numbers and geometric figures was on the curriculum. The circle, the square and the triangle have a level of meaning that cannot be accessed by the rational mind, but *can* be perceived by the feeling imagination. And the only place this dimension of geometry has survived over the centuries is in astrological theory and practice.

It's because of numbers that music sounds either harmonious or disharmonious to our ears, always affecting us emotionally and sometimes inspiring us spiritually. Similarly, perfect geometrical shapes when they are used in works of art have an uplifting effect on the beholder. So, although numbers are normally used by the rational left-brain for practical, mundane purposes, when embodied in a piece of music or expressed as architectural proportions, they can be appreciated by the intuitive right-brain and become meaningful on a soul level.[1]

So far my investigation into ancient cosmology had indicated that astrology works through the agency of universal cosmic principles, and I now began to realise how relevant the qualitative dimension of numbers was to my quest. Perhaps the principles we find personified in mythology as different gods and goddesses are in their essence numbers pure and simple. Following this idea up I turned to Plato. He should know, I thought, because like his forerunner Pythagoras he had studied in the Egyptian temples, and must have been familiar with the ancient body of knowledge that was still preserved in his time.

His *Timaeus* contains a version of the creation story which describes how the universe emerged through the power of numbers:

> When he set in order the heaven, he made this image eternal but moving according to number, while eternity itself rests in unity; and this image we call time. For there were no days and nights and months and years before the heaven was created, but when he constructed the heaven he created them also. They are all parts of time, and the past and future are created species of time, which we unconsciously but wrongly transfer to the eternal essence

> ...These are the forms of time, which imitate eternity and revolve according to a law of number.[2]

To an astrologer this last sentence sounds as if Plato is speaking of the astrological archetypes which 'revolve according to a law of number' in the sense that, within a twenty-four hour period, each of the twelve sections of the *ecliptic* (see Glossary) – the signs of the zodiac symbolising the archetypes – rise in turn on the eastern horizon.

The ancients recognised that numbers are fundamental to the organisation of the solar system, an understanding that could have arisen through their study of the geometric relationships created by sun, moon and stars. 'It is the planets and their continuing interaction in the sky that is the origin of the ancient concept of number, from counting right up to concepts of time as planetary configurations', writes Richard Heath in his book *The Matrix of Creation.*[3] Heath claims that our pre-historic ancestors, who were able to visualise numbers and geometry better than we can do so today, recognised the code of number that lies behind the harmony of creation.

From Plato's writings it's clear he believed that the phenomenal world was framed to mirror a divine eternal order, both being structured according to the same numerical laws. As he wrote in the *Epinomis:*

> The most important and first (study) is of numbers in themselves, not of those that are corporeal, but of the whole origin of the odd and the even, and the greatness of their influence on the nature of reality.[4]

And in *The Laws* he writes that geometry comes after numbers in importance, which:

> proves to be a manifest likening of numbers not like one another by nature in respect of the province of planes, and this will be clearly seen by him who is able to understand it to be a marvel, not of human but of divine origin.

Also, in *The Laws,* Plato refers to a secret canon of numbers that apply to every realm of human life, for example to government, astronomy, geometry and divination. And he goes on to mention a method of relating the different classes of phenomena to this numerical system by which, he tells us:

> the student may come to know the unifying principles in nature, which manifest in every diagram and system of number and every combination of harmony, and the agreement of the revolution of the stars.[5]

As an astrologer I immediately understood his unifying principles of nature to be identical with the astrological archetypes – the organising principles in relation to which all phenomena are classified in astrology, and which manifest in 'the agreement of the revolution of the stars'. Also ancient cosmologies and religions the world over all rest on correspondences between numbers and cosmic laws. No wonder Plato maintained that mathematics was the source of all meaning, and that it offered us knowledge of the eternal divine order from which he recommended that earthly aesthetic and moral values should be derived!

The geometry of creation

Creation appears to have created itself through the unfolding of geometry. As seen by the ancients, however, higher-dimensional geometry bridges into divinity and therefore divinity lies at the source of creation. We can discover for ourselves whether this is true experientially through the practice of meditating on mandalas, whose shapes and colours lead the eye via their geometric pathways into the very centre of the image as the source of its forms.

The term 'sacred geometry' is used of geometry's most perfect shapes which the ancients held sacred to the gods. These were shapes whose numbers were believed to emanate directly from the cosmic mind. And all creation in the macrocosm and microcosm was founded on the geometric proportions created by the relationships between their numbers. Thus the resulting figures were seen as visible expressions of the structure-giving powers of the cosmos.

According to Plato's philosophy, not only did sacred numbers originate in the divine mind but also what he called the 'Ideas'. (Our English word 'idea' is written in the same way as the Greek word it derives from, though when used in Plato's sense it's written with a capital 'I'.) The significance of Plato's use of Idea became clear to me now in the context of ancient cosmology. The structure-giving powers of the cosmos that Plato was speaking about were ideas in the literal sense of the word, because they arose within an all-embracing universal mind.

Many years ago in my philosophy degree course I remember our professor insisting that Plato's Ideas were transcendent to physical reality, which made Plato a dualist. But why shouldn't they be both immanent in *and* transcendent to the material world at the same time, I argued. They could be present and active in the material universe, while also existing in a higher, spiritual dimension, couldn't they?

My opinion of those days has been borne out since by what I've learned about the all-inclusiveness of ancient cosmology. For example, for the Egyptians Atum as the universal mind did not exist in a dimension separate from the human, but was immanent in all human minds. And, as I explained in Chapter 1, we can see our individual minds as participating in the great universal mind just as our individual psyches all participate in the *anima mundi* (see Figure 4, p. 33).

The quantum physicists' conception of a unified quantum field is reminiscent of the Egyptian conception of the mind of Atum, as we have seen. And we can also relate Plato's invisible geometric forms, which give structure to what comes into physical existence, to quantum theory. They are very like the 'probability patterns' described by physicists such as David Bohm, who envisaged these patterns as present in waves of energy within the quantum flow, and suggested that the information they carry could organise the particles formed when a wave collapses.

Geometric figures as ordering principles are found throughout the realm of Nature – for example in the hexagonal and pentagonal arrangements of flower petals. And we can actually watch geometry emerging into visibility by passing a magnet over a heap of iron filings, or by oscillating water by conducting sound waves through it.

The researcher Alexander Lauterwasser, who is continuing the work of the Swiss scientist Hans Jenny on cymatics, has videoed a variety of geometric figures emerging from water while being vibrated by sound waves. He explains how, in their initial condition, the water molecules are in a state of chaotic confusion. But when the sound waves are introduced they begin to oscillate in unison, and finally geometric shapes emerge. He found that different frequencies of vibration produce different geometry, and one of the illustrations contained in his book *Water Sound Images* is an example of an astrologically relevant dodecahedron arising by this means![6]

Since the development of systems science back in the 1980s, systems analysts have been studying the process by which mathematically based patterns emerge spontaneously from chaotic systems, fractal patterning being an example. They have reached the conclusion that states of harmonious equilibrium can arise out of initial states of chaos of their own accord, the reason being that systems as such have the capacity to self-organise. And this is what cosmologists now believe happened during the formation of our solar system.

The astronomer and astrologer Johannes Kepler (1571–1630) made the claim that the world was ordered by geometry. Announcing that his subject of study was 'the archetype of the movable world', he said he'd discovered that the solar system was constituted of Plato's five polyhedra, which defined the distances between the planets (see Figure 6). These solids had been identified by Plato as the basic building blocks of the cosmos. In contrast to other geometric shapes they were special because each

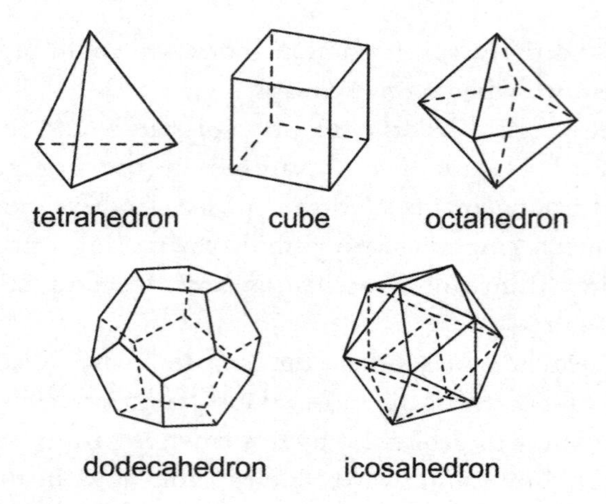

Figure 6. Plato's five polyhedra.

could be contained within a sphere with all its vertices touching the perimeter. And, as the sphere symbolised heaven for the ancients, this fact was enough to sanctify them.

Kepler devised a model in which he fitted the polyhedra neatly into the spaces between the planets – the cube between Saturn and Jupiter, the tetrahedron between Jupiter and Mars, the octahedron between Venus and Mercury, the dodecahedron between Mars and the earth, and the icosahedron between the earth and Venus.

The laws of planetary motion that he had discovered, and for which he's renowned, were of secondary importance to Kepler. He believed his life's work was to deliver material proof of the ancient concept of the harmony of the spheres, and in so doing to Christianise the pagan wisdom of ancient Egypt. 'I am free,' he wrote:

> … to give myself up to the sacred madness, I am free to taunt mortals with the frank confession that I am stealing the golden vessels of the Egyptians, in order to build of them a temple for my God far from the territory of Egypt.[7]

He didn't succeed. But from the ratios of planetary motion, and the distances between the planets that he'd calculated, Kepler derived the musical notes that he believed produced the legendary music of the spheres, whereby Saturn and Jupiter sang bass, Mars tenor, the earth and Venus alto and Mercury soprano.[8] When expressed as music the mathematical ratios introduced qualities and therefore meaning, not only in relation to the five geometric solids but also to the planets themselves.

To explain the qualities of the five polyhedra, Kepler divided them into three primary and two secondary solids; the cube, the tetrahedron and the dodecahedron being primary, and the octagon and icosahedron secondary. He went on to identify their genders and to describe the 'weddings' between them.

> There are as it were two noteworthy weddings of these figures made from different classes: the males, the cube and the dodecahedron, among the primary; the females, the octahedron and the icosahedron, among the secondary, to which is added one as it were bachelor or hermaphrodite, the tetrahedron.[9]

The five solids also had their specific qualities in ancient times. Pythagoras associated them with the five elements – the cube with earth, the tetrahedron with fire, the octahedron with air, the icosahedron with water and the dodecahedron with ether, thus symbolising their ranges of meaning. Like Kepler, Pythagoras had also maintained that the elements could be represented by musical frequencies, and had taught that each element corresponds to a particular geometric shape.

In ancient Egyptian science the elements were considered to be manifestations of the first visible states that energy adopts as it emerges as substance on the physical plane. Ether, as the subtlest and most spiritual element, corresponded to the universal life force flowing through the whole of creation ensuring its unity. (It's been suggested that this force may in fact be the 'dark energy' recently discovered by astrophysicists.) And, just as ether is present in everything, so the dodecahedron corresponding to it was said to contain the cube, the tetrahedron, and the octahedron. However it could not completely contain the icosahedron, as this Platonic solid transcended the bounds of the physical universe.

All this appeared to me to be very relevant to astrology, because the zodiac, seen as a three-dimensional solid, becomes a dodecahedron with its twelve faces corresponding to the twelve signs. In this context it was interesting to discover that Plato thought the ultimate shape of the universe must be of a dodecahedron, and recent results from microwave radiation measurements of the universe indicate he could be right!

Just as the element ether is present in everything in our universe, so the zodiac contains symbolically everything that exists. Also the qualities that Pythagoras and Kepler associated with the elements can be applied to the signs of the zodiac and their inter-relationships, which are the basis of astrology. The sides of the polyhedra, for example, correspond to the lines drawn by the main *aspects* in a chart. *Squares* are then the sides of cubes, for example, and *trines* the sides of tetrahedrons (see Glossary). And, as we shall see in Chapter 8, not only does the zodiac have an absolute internal geometry, but an individual chart has its own particular geometry created by the aspect patterns within it.

The ancients believed that, when they were related to one another in geometric figures, numbers became powerful, and that the energies thus generated could be used beneficially. The truth of this belief is borne out by a number of alternative healing methods. The geometry that structures crystals, for example, is used in crystal healing to harmonise the subtle bodies and encourage their coherence.[10] And meditation on geometric shapes is a practice that has been used in spiritual schools since ancient times to harmonise and transform states of mind.

Similarly the presence of perfect geometric proportions in a temple or church creates a spiritually beneficial environment. In Egypt not only the architecture of a temple but the proportions of the art works it contained – the statues and engraved reliefs – were all designed to conform to the aesthetic canon in this respect.

Knowledge of these secrets was part of the ancient wisdom that the crusaders recovered in the eleventh and twelfth centuries from the Near East where it had been preserved. And the magnificent European Gothic cathedrals were the result.

The spiritual potential of sacred geometry is also indicated by the belief prevalent in ancient cultures that the statue of a god, when carved with perfect proportions, could become animated by the divinity it represented, because material perfection attracted spiritual powers to manifest in the physical world through resonant empathy.

Most holy tetraktys

A mathematical demonstration of the evolution of the cosmos was on the syllabus for all maths students in the Egyptian temple academies. And the tetraktys, which is a diagram of the first ten digits arranged in pyramid form, was a teaching aid used for this purpose (see Figure 7). Although Pythagoras has taken the credit for inventing this geometric allegory, it is very likely to have originated in Egypt. Pythagoras studied in the Egyptian temples for twenty-five years before founding his own academy on the Egyptian model, and Egypt is where he learned his geometry.[11]

In the tetraktys we see the first ten digits arranged as an isosceles triangle, which points to the significance and potency of

Figure 7. The holy tetraktys.

the triangle shape as it appears in pyramids. This simple diagram shows the hierarchy of the numbers governing the universe, and gives a visible demonstration of their eternal mathematical relationships. It can be seen both as a static geometric shape, and also as a numerical sequence.

Each of its ten digits represents an archetypal power of the highest order. From their timeless spiritual dimension they manifest in our space-time world to create empirical existence, and give meaning to all its phenomena while remaining at the same time transcendent to them.

The tetraktys was revered as holy in Greece because it was believed to contain the secret of life. And those to whom its mysteries were revealed were first required to take an oath that invoked the following blessing:

> Bless us Divine Number, generating gods and men, most holy tetraktys, containing root and source of ever-flowing creation. For divine number begins in profound pure unity, and coming to the holy four, begets the mother of all. The all encompassing, the first-born, never swerving nor tiring, the holy ten, keyholder of all.[12]

The tetraktys not only demonstrates the mathematical basis of creation, but the order of the ten digits in pyramid form, together with their inter-relationships, reveals creation's first steps. These symbolise not only a sequence of events that happened billions of years ago at the birth of the universe; they also represent an ongoing process of descent from unity into multiplicity – present as well as past. We are descending into multiplicity all the time, for example through creating the world with each act of perception.

At the top is One standing for the absolute, the all. Below it is Two, which is where the primary schism occurs. The positive and negative cosmic forces polarise here to create the dualities of male and female, light and dark, day and night etc. On this second rung, therefore, we are continuously see-sawing between the polar opposites.

Philosophically the primary schism occurs when the creative consciousness becomes conscious of itself, thus creating the subject-object polarity of Self and Other. In other words the creation story begins with a fall in consciousness from unity to duality; and, conversely, the goal of all spiritual paths is to reverse this and return consciousness to its original condition of oneness.

The re-unification of Self and Other would be demonstrated geometrically by Plato. He would start with two circles as in Figure 8. Circle A he called 'the Same' and circle B 'the Other'. A, he said, represented the positive masculine and B the negative feminine forces. Next he would demonstrate how they could be united, as in the following figure, through the creation of a *vesica piscis* (C). In the *vesica* the duality of the Same and the Other is symbolically overcome, and both are united in a *unio mystica* or mystic marriage.

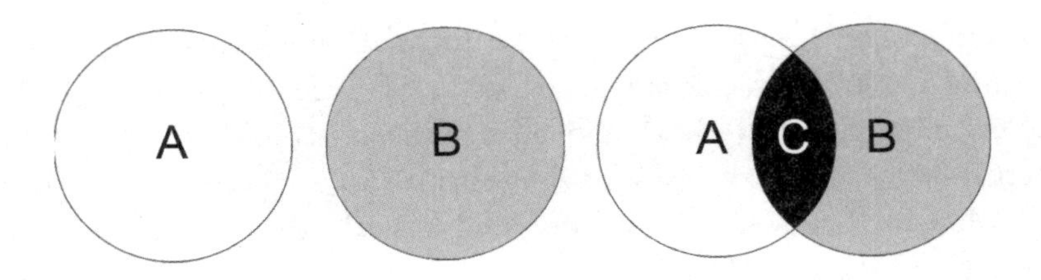

Figure 8. The creation of a vesica piscis.

The next line down in the tetraktys has three digits. Three is expressed spatially by the triangle or tetrahedron. A 'capstone' has already been created for the tetraktys pyramid by the digits in lines one and two, and this is now expressed numerically in line three. Three resolves the tension of the opposition of the two forces by reconciling them through the mediation of a third transcendent point lying above and between them.

Like many other elements of Christian theology, the Christian trinity of 'three in one and one in three' was a concept derived from Egypt. There it was taught that: 'Three are all the gods,

Amun, Ra, Ptah; there are none like them. Hidden in his name as Amun he is Ra and his body is Ptah.' In other words Amun (a version of Atum) was the first creator god or the One. And out of him, as he unfolded like a fan, emerged the two other main creators, Ra and Ptah.

Three is also the number of the astrological modalities – the *cardinal, fixed* and *mutable* modes of energy, words which describe how energies move (see Glossary). And, as there can be no motion outside time, with number Three the time factor enters into creation.

There follows number Four. Up to this point the process of creation has taken place purely in metaphysical dimensions, but with Four matter and space emerge. The triangle becomes a square, or three-dimensionally a cube – the Platonic solid that represents earth. And the four base-line digits of the tetraktys come to symbolise the elements of earth, water, air and fire, which are the first forms adopted when matter emerges either in a solid, liquid, gaseous or plasmic form. Empedocles who lived around 450 BC has been accredited with the discovery of the four elements, but these had already been around in Egyptian cosmology for centuries – maybe for millennia – before his birth.

The association of the polyhedra with the elements is a further stage in the evolution of abstract numbers into material forms with qualities that can be sensed. Plato explained that transformations of energy from one element into another in the physical world, such as water freezing into ice and becoming solid, or gases igniting and becoming fire, are all processes whereby the elemental geometric solids get broken up and then reform. And, as we shall see in Chapter 8, something similar to these transformations also happens when astrological archetypes merge to create new combinations of meaning..

With the number Four the four cardinal points of time and space are established which provide a basis for material creation. These are the peaks in the solar year of the two solstices and the two equinoxes together with the four cardinal directions of north, south, east and west. Thus the base-line digits in the tetraktys manifest, for example, in the form of the gods of the four directions. These have been given different names in different cultures, but essentially they are always manifestations of the same numbers, and, as we will see, also represent the same four astrological archetypes.

The potency of twelve

The tetraktys is said to contain the whole of creation, because from the primary ten digits, corresponding to our ten fingers, all the other numbers in the universe can be derived. As is written in the blessing: 'For divine number begins in profound pure unity, and coming to the holy four, begets the mother of all'. And this begetting comes about because, connected through their numerical relationships, the ten digits can be multiplied and added together infinitely. So a fifth step, for example, would be to multiply the third line by the base line of the tetraktys to engender the number Twelve (3 x 4 = 12).

Twelve was seen as the most potent of the numbers in our space-time world. It is pre-eminent in the vision of the holy city of Jerusalem in the Biblical book of Revelation, which is an image of the cosmos structured according to divine mathematical laws.[13] It manifests in the dodecahedron, which Plato said was the ultimate shape of the universe. Twelve is the sum of its faces, and as they are pentagonal that means there are sixty sides in all. As we shall see Twelve and Sixty are key numbers in the structuring of time and space.

The Egyptians are credited with being the first to divide the day (and night) into twelve hours of sixty minutes, and the Babylonians with dividing the circle into 360 degrees. 360 is a number containing many factors. Two, three, four, six, twelve, thirty, sixty and one hundred and twenty are all included in it, which points to a logic behind the conventional number of degrees in circles, and to a deeper significance behind the division of the zodiac into twelve thirty-degree sections.

The pyramid as a shape is a visible demonstration of the principle of Twelve. Taking the Great Pyramid as an example, it has a square base representing the fourfold principle of matter, and corners perfectly oriented to the four directions. Then it has four isosceles triangles arising from this base to meet at the apex, with their sides representing the three-fold modes of energy. Thus the pyramid can be seen as a geometric manifestation of the number Twelve and as such was immeasurably potent.

The power of twelve was used by the Egyptians in the designs of their temple halls which often include twelve columns. Twelve also structured the Egyptian calendar in which there were twelve months in the year, each named after a divinity whose feast days were celebrated in turn as their month came round. In addition every hour of the twelve-hour day had its special qualities to which the astrologers would pay careful attention.

All of this sheds light on the origin of the qualities attributed in astrology to the twelve signs of the zodiac. For example it's likely that their modern-day meanings have a history that goes back to the qualities associated with the personalities of the ruling divinities of the twelve months in ancient Egyptian mythology.

The conception of One becoming Twelve is dramatised in the Greek myth of the divine child being killed and eaten by twelve Titans, which, we learn, so disgusted Zeus that he blasted the Titans and created mankind from their ashes. The moral of the story is that the same basic twelve archetypes are present in us all, just as we each have all the zodiac signs in our birth charts. A similar narrative can be found in the myths of Osiris and Dionysius, who were both dismembered and whose body parts were widely dispersed, only to be later collected and miraculously put back together again.

And then there is Jesus whose body was consumed at the Last Supper by his twelve disciples, each of whom traditionally represents an astrological archetype – as Leonardo da Vinci well knew when he painted his fresco *The Last Supper*. Before starting work he is reported to have combed the streets of Milan in search of twelve suitable models to sit for him, whose physiognomies expressed the traits that are typical of each zodiac sign.

There were also twelve ordeals that an initiate in the ancient mystery schools needed to pass through. Like Hercules accomplishing his twelve labours (a myth that is likely to have been a mystery school allegory) the initiate had to pass twelve tests in order to become enlightened.[14] And possibly there's a link between the myth of the hero accomplishing his twelve labours, found in Egyptian and Babylonian as well as Greek mythology, and the astronomical cycle of the *precession* (see Glossary). In this cycle the point of the vernal equinox moves through twelve 2160-year phases following the zodiac constellations in reverse order.

Although writers of histories of astronomy go on attributing the discovery of precession to the Greek mathematician Hipparchus (190–120 BC), reluctant to admit that the civilisations preceding the Greek were in any way astronomically clued up, Rosemary Clark makes a case for the precessional cycle being known from the earliest Egyptian dynasties onwards, and given religious significance.[15]

If, instead of multiplying the digits from the two bottom lines of the tetraktys, we add them together we get seven. Seven has from time immemorial been invested with magical significance. It may have been seen as a mystical number because it has no factors (meaning nothing divides into it). It was also considered the sum of heaven and earth, and of spirit and matter which are symbolised by the Three and Four respectively. There were also seven veils to be lifted on the spiritual path before Truth could be revealed, and seven steps on the mystical ladder leading from earth to heaven. Also in traditional astrology there were seven planets, which included the sun and moon and the five planets visible to the naked eye.

The numbers of creation

The Egyptian creation myths are complex as there were four main theological schools in ancient Egypt each with their different versions of the stories. These were at Heliopolis, Memphis, Hermopolis and Thebes. However rather than being separate, self-contained mythic traditions, as they are seen by many Egyptologists, Clark believes they describe different cycles of creation within one unified cosmological vision.[16] She writes:

> Each cosmogony was present in the formative years of Egyptian civilisation, but at the same time each creation myth, no matter how seemingly divided from the others by time or place in popularity, formed an integral element in an overall worldview which the Egyptians consciously understood in the large context of time.[17]

The Egyptian creation myths are interesting because they are probably older than the familiar story told in Genesis, and parallels can be more clearly drawn between their narrative elements and the mathematical steps of the tetraktys. Like the tetraktys they describe a descent of divine creative power from unity into multiplicity through the creation of a series of worlds. As I understand the symbolism of the myths, the first three worlds to emerge were all in higher dimensions, and only with the fourth world, symbolised by the base line of the tetraktys, did our space-time world appear.

In the creation myth of Memphis Ptah, the self-engendered creator god first imagines all the phenomena of creation in his heart, and then manifests the universe by speaking his thoughts out loud.[18] On the other hand the Heliopolitan creation story describes the whole action happening in the mind of Atum, who we meet in the beginning lying supine in the waters of Nun, 'the darkness, the nothing'.

The first matter appeared as a mound which rose out of the waters, and Atum placed a stone on it so the chaotic waters could not overwhelm the dry land. This was the original *benben* stone, an omphalos like the foundation rock at Jerusalem that held down the waters of the abyss, allowing civilisation to develop. Upon this first mound the temple of Heliopolis was constructed with the benben stone at the centre of its innermost sanctuary.

The Heliopolitan creation story continues with Atum spitting out the dual cosmic principles Shu (air) and Tefnut (moisture) by sneezing (or in some versions by masturbating). They engender Nut (the sky) and Geb (the earth) who then give birth to the four archetypes that rule the human and natural life cycles – Osiris, Isis, Set and Nepthys. Together these powers make up the hierarchy of nine cosmic principles known as the Ennead (see Figure 10, p. 82). Then the trinity of principles behind transformational processes (the Dendera triad) were added to them to make twelve in all.[19]

Perhaps the Hermopolitan creation myth is the earliest one, as, according to Clark, its sources could lie in the archaic pre-dynastic period of Egyptian history. Here the birth of the macrocosm from the primeval waters of Nun is triggered when Nun's female emanation, Naunet, separates from him. This primordial couple then procreate to produce the Ogdoad, a family of eight primary archetypes.

Each principle had its number and its name in Egyptian cosmogony, and in the sounding of its name lay its power. Thus Thoth, an emanation of Atum, continued the process of creation by articulating a series of divine words, each of which resonating at a specific frequency. This caused matter to emerge from the invisible dimensions and manifest on the physical plane.

At the same time his consort Maat, who was in charge of the matrices, dispensed the geometric patterns that structure physical forms.[20] And thus primeval chaos was ordered. The Egyptian temple choirs helped to maintain this order by singing hymns to Maat composed of the basic harmonies of creation, said to echo the perpetual music of the spheres.

And this is not all a fiction; apparently there's something in it. Hans Jenny (see p. 61), who studied the way wave forms influence matter, discovered that when the letter 'O' was pronounced aloud it produced a perfectly spherical vibration in the air, and that the other letters of the alphabet when sounded also produced characteristic patterns.[21]

In ancient Egypt it was believed that a society would only flourish if it conformed to the laws of proportion found in Nature. The numbers that generated these proportions were seen as active forces in the cosmos, and their patterns arose through their interactions. For example the logarithmic spiral generated when *phi* is the key number, which governs the pattern of leaves emerging spirally round the stems of plants, as well as the spiral growth of goats' horns and snail shells, is found everywhere in Egyptian art. Additionally artefacts from the earliest dynasties onwards demonstrate an awareness of the ratios of the Golden Section, which Egyptian artists would use to regulate the proportions and forms in their art and architecture.

The ancient art of gematria, which is discussed by Plato in the *Cratylos,* according to which numerical values were assigned to the letters of the alphabet, also appeared relevant to my investigation into the qualitative dimension of numbers, and so I looked into it. It appears that the association of the basic sounds of speech with numbers could go back to a time well before the invention of reading and writing. Numeracy in any case is believed to be more ancient than literacy. Gematria also assigned numerical values to words and phrases. And Plato, who believed that the power of a word lies in its number, describes how, within this system, letters, words and phrases can be substituted for others having the same numerical values.

An adept versed in gematria could identify the archetypal principle a god represented through the numerical value of the letters that made up his name. The gods all had their specific numbers and could be recognised by them, which was useful because, whereas their names varied in different languages, their numbers were everywhere the same.

To conclude this present section, I'd discovered that numbers are far more than a mere human invention. As cosmic forces they're fundamental to the structure of the cosmos. Also, in the vein of Pythagorean thinking, they can reveal the otherwise inaccessible nature of deeper cosmic truths. And also – very important – numbers have a qualitative dimension. As number symbolism is basic to astrology, we can describe astrology as the art that reveals the numbers governing the universe. Finally the qualities of the geometric shapes investigated by Plato and Kepler, which they described using the symbols of the five elements, now appeared to me as keys to hidden doors that could possibly lead me to the very sources of astrological meaning.

CHAPTER 4

All About Archetypes

As seen by the Greeks

There were further steps I now had to take on my quest to find out how astrology works. For example I'd discovered that archetypes are like the skeleton from which all meaning in the cosmos is fleshed out. They are the bare structural bones on which life is hung, so what exactly they are now needed to be clarified. My first move was to the dictionary where I learned that 'archetype' stems from the Greek word *archetypon* meaning 'the first mould'. From this I concluded that archetypes must be the originals from which all later copies were taken – in other words they were the first edition of life and therefore should be the most perfect.

As we have seen, Pythagoras and Plato were of the opinion that all things in the universe derive from abstract numbers. And archetypes for them were numbers which, when linked together in sequences, acted as the formative causes and regulators of everything in the phenomenal world. They were on the same elevated plane as the Ideas, which for Plato were archetypes in the sense that they underpinned material phenomena. I saw his Ideas as eternal forms shining out like the veins of a leaf from deep within the fabric of creation.

As Plato saw it an individual hawk was cast in the mould of the archetypal Idea of a hawk, and, whereas the flesh and blood bird is born and dies, the archetypal Idea lives on eternally. Also Plato would have considered a woman beautiful to the extent that she participated in the archetype of beauty. So, when a man fell in love with her, he was surrendering not to the human love object, but to the goddess Aphrodite manifesting through her.

The Neo-Platonists believed that, when the Ideas of love and beauty are embodied in a woman beautiful in body and soul, she becomes a fit vessel for the goddess who in a sense can then inhabit her. Also they saw the archetypes as responsible for the personality differences between the gods. For example the warrior god Ares was known to be brutally uncouth, while Aphrodite was sweetly seductive; seen archetypally these two are polar opposites.

In Homer's era the realm of the gods impinged on the human world. Their awesome presence was felt in nature, and the course of human destiny lay in their hands. However, prayers and offerings could influence their decisions. The Greek deities may have had absolute power over the domains they ruled, but a sacrifice to Poseidon could calm a storm and ensure a safe sea passage. It was believed in those days that humans had a degree of power to affect the unfolding purposes of the gods. And this was because the human will and the divine will were not considered as essentially separate. If human willpower was employed consciously, and backed by a firm purpose, 'co-creativity' with the gods could occur.

Within the Greek cosmos the gods represented a higher order of intelligence with an overview of things that was denied to man. However they also appeared very human. Their different character traits reflect the strengths and weaknesses of human personalities on a blown-up scale, and their adventures follow the human life patterns that repeat in every generation.

These parallels allow an individual's experiences to be interpreted within the wider context of meaning offered by the myths of the gods, without proving that the gods were archetypes – at least not archetypes of the eternal sort. However the archetypes *per se* could manifest in the guise of gods, creating a mediating level between human life and the higher spiritual planes.

It was when the Egyptian culture merged with the Greek that the Egyptian Neters, described in the following section, became known as the 'Archai'. These, like the Neters, had archetypal status and were honoured, for example, as 'the twelve gods' of Athens. Offerings were made on their altar to mark the start of the annual festival known as the Dionysia, whose roots also lay in Egypt. It derived from the ancient annual festivals in the Egyptian temples at which sacred dramas were performed that recreated the passion, death and resurrection of Osiris.

The Egyptian Neters

The Egyptians called the archetypes Neters *(Neteru* is the Egyptian plural). These were understood as eternal powers that were both immanent in and transcendent to the material universe. Although they could appear in the form of gods and goddesses, in their essence they were as pure and bare as stars (see Figure 9).

The Egyptian word *Neter* is generally translated as 'god', but I believe a distinction should be made here as Neters are much more like mathematical principles. We saw in Chapter 3 how the universe came into being in Egyptian creation myths through the impetus given by cosmic principles exercising their powers through numbers. The Neters are these creative principles, while the gods are a rung below them in the cosmic hierarchy. They come and go but the Neters remain for ever and ever.

However as Osiris and Isis, who have Neter status, are represented by statues and praised as gods in hymns, the distinction between Neters and gods can get somewhat blurred. I concluded that some Neters can take the form of gods, but that

Figure 9. The Egyptian Neters.

not all gods are Neters. According to Rosemary Clark, there are altogether forty-two Neters in Egyptian theology, of which twelve are significant for life on earth:

> In this system, the emphasis is on human and divine life as an indivisible energy dynamic, acted out and mirrored thorough all phases of the natural world. Nature is the supreme model in this worldview, and every concept in the system employs a natural element that articulates a spiritual reality in a concrete, obvious way.[1]

This helps us understand why the Egyptian gods are portrayed with animal body parts, a custom that greatly offended the Christian missionaries. It was to represent the archetypal powers they possess that lie beyond the human. The hawk-headed god Horus, for example, possessed the broad, superior vision of this bird of prey, and the lion-headed goddess Sekhmet the power, ferociousness and majesty of the king of the beasts. The Egyptians were coming from a belief in the oneness of life, and saw the powers specific to certain animals as present to a greater or lesser extent in everything and everybody. And, by the way, the zodiac animals should also be understood in this vein.

The budding astrologers who studied in the Egyptian temple academies were required to understand the natures of the twelve Neters, as this knowledge was central to their training in astrology. An Egyptian astrologer was not only expected to be familiar with their different qualities, but also had to know their order of ascendancy within time cycles. As astrology deals with the qualitative dimension of maths observable in the moving geometry of the sky, mathematics and astronomy were also compulsory subjects on the syllabus.

The annual cycle of the sun was divided by the lunar cycle into twelve months represented by the month deities. Each had a number and a name which served to describe their particular nature. Additionally every hour of the day had its own number and quality. And the same succession of twelve Neters determined the order of the qualitatively different time periods within both the diurnal and the annual sun cycles. Traditionally there were favourable or unfavourable days and hours, which depended on the particular combination of the energies of the Neters ruling the hour, the day and the month.

The alternative Egyptologist Schwaller de Lubicz (1887–1961) writes:

> In every month of each season of the year, every hour of the day has its Neter, because each one of these hours has its own character … Purely material reasons no longer explain why the season, even the month and the precise date, must be taken into account for the best results. Invisible cosmic influences come into play.[2]

Schwaller's understanding of Egypt, gathered from the decades he spent measuring the dimensions of the Egyptian temples, diverges from that of the more orthodox Egyptologists. Their reductionist approach keeps them narrowly focused on concrete facts while he gave a symbolist's account of Egyptian purpose and intention.

For example in the proportions of the Temple of Luxor he saw the laws reflected that govern the cosmos, and found them identical to those that determine the proportions of the human body.[3] Schwaller saw the art and architecture that he describes in his books as didactic. He believes the Egyptians used these symbolic means to convey the ancient wisdom they possessed to future generations – a wisdom that was at the same time both rigorously scientific and deeply spiritual.

According to Herodotus, who visited the temples of Heliopolis and Thebes in the fifth century BC and conversed there with the temple priests, 'The Egyptians first brought into use the names of the twelve gods, which the Greeks took over from them'.[4] He also reported: 'They say that 17,000 years before the reign of Amasis the twelve gods were produced from the eight'.[5] Also that 'the names of nearly all the gods came to Greece from Egypt …

for the names of all the gods have been known in Egypt from the beginning of time'.[6]

His testimony points to the Greeks adopting the creator gods called the Archai from Egypt where, as mentioned above, knowledge of these powers, called the Neteru by the Egyptians, had a very long history. The 'eight' referred to above are the eight Neters of the Hermopolitan Ogdoad (see p. 73). The timescale suggested by Herodotus, which is taken with a pinch of salt by most of his commentators, supports the theory that Egyptian culture was a legacy from that vanished advanced civilisation of prehistory mentioned earlier.

We learn from Herodotus that:

> The Egyptians were also the first to assign each month and each day to a particular deity, and to foretell by the date of a man's birth his character, his fortunes, and the day of his death – a discovery which Greek poets have turned to account.[7]

This important statement overturns the widespread opinion that the Egyptians learned their astrology from the Greeks. Instead it is clear that the Greeks inherited astrology together with their knowledge of the archetypal principles from the Egyptians and much more. Herodotus also states that Egypt is the source of the Dionysian mystery tradition as well as being the inspiration behind the great Greek religious festivals.

The Egyptian astrologer's knowledge of the natures of the individual Neters, and of the times when each would become dominant within time cycles, allowed him to anticipate the conditions likely to prevail during any time period. Also, as he could foresee the kind of ideas and events that were about to arise, he was in a position to warn the rulers of them. The Egyptians were aware that a social order needs to respond sensitively to change in order remain in balance and flourish. And I suggest it was thanks to their astrologers that the Egyptian civilisation managed to endure, with ups and downs, for more than three thousand years!

The hierarchy of the archetypes

As the description of the tetraktys in Chapter 3 has shown, archetypes are numerical powers that can be arranged in a hierarchy, and which evolve from one another in stages. For example their step by step emergence as creative powers can be followed in the Heliopolitan creation myth.

At the top of the pile in Figure 10 is Atum, the One and chronologically the first Neter, who begets the Two, Shu and Tefnut, thus creating the trinity of Three by which Nut and Geb are engendered. They then become the procreators of the Four represented by the Neters Osiris, Isis, Set and Nepthys. By multiplying the Four by the Three Twelve is attained – which is the sum of all the archetypes governing our space-time world.

Except for Atum, who contains both genders, this pantheon is composed of male-female pairs, the division into positive (male) and negative (female) being the primordial energy distinction entering with the primary schism. It's the first qualitative distinction of the Neters. Below the Neters in the Egyptian hierarchy were ranked a multitude of lesser gods and spiritual powers, whereas further down there extended a whole range of intelligences, elementals and nature spirits.

A liturgy of rites and observances was performed at the main Egyptian temples celebrating each Neter in turn during the course of the year. This annual cycle was called the 'Iru', and its

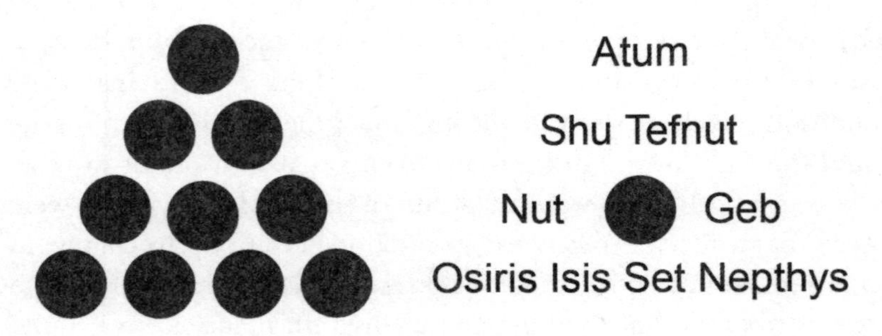

Figure 10. The tetraktys and the hierarchy of the Neters.

ceremonies were adapted to honour the specific powers of each Neter. I suggest that the origins of the different meanings of the twelve signs in astrology could be traced back to this practice. According to Rosemary Clark, who became my authority on Egyptian sacred science, the twelve Neters correspond to the astrological archetypes as follows (the Greek names are used instead of the Egyptian where these are better known): Osiris = Aries, Hathor = Taurus, Thoth = Gemini, Nepthys = Cancer, Horus = Leo, Geb = Virgo, Isis = Libra, Anpu = Scorpio, Sekhmet = Sagittarius, Set = Capricorn, Maat = Aquarius and Nut = Pisces.[8]

The understanding of the order of importance of the Neters within the hierarchy that is offered by the symbol of the tetraktys pyramid will be relevant to my discussion of the meanings of the astrological archetypes later in the book, and to my suggestions on how they should be weighted.

Jung and archetypes

As we have seen, human awareness of the presence of archetypal principles as formative powers in the cosmos goes back to very ancient times. However, when Carl Jung started delving into them in the early part of the twentieth century, they had long been neglected and were largely forgotten. In developing an interest in them Jung was picking up on a very ancient thread, and it led him back into the archaic view of the cosmos.

He first came across the archetypes in his work as an analytical psychologist when he was interpreting his patients' dreams. He noticed that the images and narratives that emerged in them were similar to those found in myths and fairy tales the world over. They included symbols of great potency that were emotionally highly charged, and which were impossible to grasp by the intellect alone. The conclusion he came to was that these 'living psychic forces' were archetypes arising from the deep ground of the collective unconscious.

Although in his earlier writings Jung describes the archetypes as psychological phenomena, later on in his career, when he was straddling the mind-matter duality, he began to see them as autonomous cosmic principles. Not only did they function as principles of order in the human psyche, preconditioning our perceptions and patterning our actions, but they were also part and parcel of the fabric of the cosmos. He indicated their geometric basis when he compared them to the axial structuring patterns within crystals.

Jung called his archetypes of the collective unconscious 'creative ordering factors' (like the ordering factors of geometry), and argued that they give a dynamic thematic structure to human experience. As he moved deeper into the archaic worldview, he fastened on the ancient concept of the *Unus Mundus* to represent the oneness of existence. However when he began to speak of the universe as a living creature, this startled the scientific establishment of his day, who could only see a heap of dead matter. Nevertheless, in the meantime, the conception of a living, intelligent cosmos has become more mainstream – at least in green and New Age circles.

In order to be understood by his contemporaries, Jung employed a variety of terms to describe the archetypes – 'gods, patterns of behaviour, conditioning forces, primordial images, unconscious dominants, organising forms, formative principles, instinctual powers, dynamisms – to give but a few examples'.[9] Le Grice, from whom this quote derives, draws on Jung's later opinions on the nature of the psyche to support his own cosmological interpretation of the archetypes, and endorse the central idea of *The Archetypal Cosmos* – that the structural order of the psyche and the structural order of the cosmos mirror each other, both ultimately resting on the same underlying ground.[10]

In these lands of darkness

In 1911 Jung took up astrology as part of his research into the structure of the psyche. 'At the moment I am looking into astrology,' he wrote, 'which seems indispensable for a proper understanding of mythology. There are strange and wondrous things in these lands of darkness.'[11]

He became skilled at using birth charts in his psycho-analytic work.[12] Parallels between the archetypes, which he defined as forces in the unconscious, and the astrological principles soon became apparent to him. He would have noticed, for example, the correspondence between the concepts associated in astrology with the archetype of Leo – such as king, gold, and lion – and the themes he had assigned to his archetype of the Hero.

When Jung describes archetypes it sounds to an astrologer as if he's talking about the astrological signs and planets. And back in the 1970s the astrologer Liz Greene, who is also a Jungian analyst, pioneered an approach by which she blended astrology with Jungian depth psychology. She took examples from Jung's wide collection of myths, fairy tales and dreams to amplify the meaning of the astrological symbols. And she also adopted structures from his theory, for example the fourfold pattern of the functions of consciousness, which she used to explain the astrological meanings of the four elements.[13] It was through Liz Greene's books that Jung's conception of archetypes came to inspire the development of psychological astrology during the second half of the twentieth century.

Although back in the 1970s my gateway into the temple of astrology had been through the portal of psychological astrology, I later realised that what this approach lacked was the cosmic dimension. Psychological astrology was after all derived from the humanist tradition. But later in his life Jung had tapped into the cosmic level of meaning of the archetypes, which had led him into the realm of the sacred where astrology has its source. Accordingly he came to see them as much more than psychological forces. They were the dynamic powers that patterned ongoing cosmic processes, and everything occurring in the cosmos could be seen in the context of their unfolding patterns, including the most trivial events of our personal lives.

The key lay in the relationship between the human psyche and the cosmos, a point that was recognised by the archetypal psychologist James Hillman in the 1980s. He built on the foundations of the archetypal theory laid down by Jung, and took things a step further. Recognising that the discipline of psychology was urgently in need of a cosmological framework, he investigated the relationship between archetypal psychology and what Jung had identified as the deeper cosmological ground. The result was the development of a new school of psychology – archetypal depth psychology.

Hillman follows Jung in seeing the archetypes as principles inherent in what we can understand as a universal mind. They act as structuring principles both in the physical world and in the human psyche, creating a matrix of meaning with universal scope. He writes:

> The soul cannot be except in one of their patterns … there is no place without gods and no activity that does not enact them. Every fantasy, every experience has its archetypal reason.[14]

The psyche is not in us; we are in the psyche

With the advent of the new millennium, the time had come to re-visit astrology in the context of postmodern thought, and the philosopher and astrologer Richard Tarnas took up the baton. In his earlier seminal work, *The Passion of the Western Mind* he had examined the stages in the evolution of the archetypal perspective in Western thought.[15] Now in *Cosmos and Psyche,* published in 2006, he turned his spotlight on the astrological archetypes, and presented the results of his thirty-year research project into how archetypal patterns manifest in history.

In accord with Jung's later thinking, Tarnas sees the archetypes as cosmic ordering principles that are both interior in and exterior to the human psyche. All-embracing in their scope, they underlie evolution *per se* and govern the course of human history. He describes them as:

> the enduring, structurally decisive essences of meaning that underlie the flux of phenomena. Grounded in the nature of the cosmos, within a universal substrate, they can be seen as ideas within a universal mind.[16]

Just as Jung had come to see the archetypes as more than psychological principles, and in his later writings described them as cosmological in scope, Tarnas now suggests that the meanings astrology assigns to the planets originate in an autonomous cosmic reality. 'I have become fully persuaded,' he writes, 'that these archetypal categories are not merely constructed but are in some sense both psychological and cosmological in nature'.[17] He suggests that their source lies in the common ground of the cosmic unconscious. The psyche is not in us; we are in the psyche, he proclaims, and thus we share in the archetypal patterning contained within the *anima mundi.*

If there is one collective psyche in which we all share, and one set of archetypal principles underlying both psyche and cosmos, this has repercussions for astrology. Then the meanings of events, which are astrology's subject matter, are not constructed by the human imagination and projected onto the world, but in a deep sense they are intrinsic to the organisation of the universe. It also implies that the symbol system used in astrology draws on these archetypes inherent in the common cosmic ground, which explains how the same archetypes appear in a similar patterning within different ancient myths and cosmogonies.

By giving provable examples of correlations between the alignments formed by the orbiting planets and events on earth, Tarnas has brought astrology down to the level of verifiable fact. In *Cosmos and Psyche* he concentrates on the cycles of the outer planets, showing how they become periodically activated during the successive phases of planetary cycles, and unfold their meanings over time. Together they are found to structure events both in the human mind and on the world stage, and point to a vast order of meaning transcending the human perspective.

Building on these foundations, Keiron Le Grice's 2010

book *The Archetypal Cosmos* investigates the implications of an archetypal cosmology for the emerging postmodern worldview. He sees his work as part of the larger academic discipline of archetypal cosmology presently being developed by scholars and researchers in California.[18] This group is studying the correlations between planetary alignments and archetypal patterns in human experience with a view to acquiring a deeper understanding of their theoretical basis, and the implications for the new cosmology.

Both Le Grice and Tarnas stress that the universe is not a machine; it is an alive, conscious organism. And, just as an organism has its internal order and coherence, so there's also an inbuilt order and coherence in the universe. This allows for stability and the repetition of the familiar although all its systems are in continuous flux. So, instead of being an object fixed for all time, the universe is open-ended – a creative process that includes scope for continual innovation.

This means that, although the astrological archetypes are eternal formative principles, how they manifest is going to vary, which is why Tarnas emphasises that astrology should never be seen as rigidly deterministic in its pronouncements. He coined the expression 'archetypally predictive' to describe the blend of continuity and innovative freedom that characterises the way the astrological archetypes manifest. And, as in *Cosmos and Psyche* we follow the cycles of the outer planets through history, and consider the many variations on their core themes that they've manifested, we are simply staggered by the extent of their creativity!

Teilhard de Chardin is quoted as saying, 'Co-extensive with their Without, there is a Within to things'.[19] And Le Grice echoes this truth when he maintains that all phenomena have not only an exterior but also an interior aspect. In other words interiority is not something exclusive to human beings as we imagine, but is present in the universe as a whole. And the reason planetary patterns are meaningful is because the solar system itself possesses such an interior dimension.

At this point I realised I'd reached the end of the road on my quest to discover how astrology works. If I put together all the understandings I'd gathered on the way, they added up to a satisfactory answer – at least for the present. However I still needed more clarity about the sources of astrological meaning, and the processes by which meaning arises through the interrelating of the components of an astrological chart. Therefore in Part Two of this book I go in search of the origins of the qualities ascribed in astrology to the astrological archetypes. The focus will shift to exploring how the meanings of these principles are expressed in a chart through geometry and the symbolism of its components. I'll also describe an archetypal approach to chart interpretation, giving a concrete example of how this works.

PART II

An Archetypal Approach to Chart Interpretation

CHAPTER 5

Horoscope Geometry

A circus of little animals

Ancient people, as we have seen, believed in the existence of a universal mind containing the numbers of the fundamental forces of creation. During my quest to understand how astrology works I had come to identify these forces with the astrological archetypes, which can be described as twelve foundational cosmic principles in the ground of the universal psyche.

Archetypal astrology is an approach to astrological chart interpretation that is based on this cosmological view. The meanings of the chart factors such as zodiac signs, houses and planets are seen to derive from the twelve basic categories of meaning associated with the astrological archetypes. These fundamental cosmic principles and their inter-relationships are symbolised in the geometry of the zodiac – a circle divided into twelve qualitatively different segments.

Looking anew at the zodiac in this context, I came to the conclusion that it was a geometric matrix and belonged to the same order of things as the hexagram that organises the particles in the gas clouds round Saturn's North Pole, or the star tetrahedron that structures water crystals. However, whereas these are examples of geometric matrices organising physical molecules, the zodiac organises the contents of the cosmic psyche, and by implication also organises the contents of our individual psyches that fractally correspond to it.

Archetypal astrology differs from other forms of astrology in being 'top down'. In other words the astrologer approaches a chart from the top, from the twelve astrological archetypes as transcendent principles, and moves down from the upper hierarchical levels that contain the overall meaning, to particular chart components, which are then understood within the context of meaning that the overarching framework provides.

The opposite takes place in a 'bottom up' interpretation. Single chart factors are seen as building blocks, which when added together produce general statements of meaning, a procedure that mirrors the way a reductionist scientist handles facts.

Zipporah Dobyns' *Expanding Astrology's Universe,* published in 1983, is an unusual astrology book in that it follows a top-down approach.[1] She describes a method of interpretation she calls the 'alphabet system' whereby the multitude of chart factors used in astrology are reduced to twelve fundamental principles. However Dobyns is basically a psychological astrologer and does not pursue the cosmological implications of her point of view.

The top-down approach not only to astrology but to life is very ancient indeed, as we've seen from the Egyptian creation myths. Their content points to the earliest forms of astrology being 'archetypal', as they would have developed in societies with top-down cosmologies. The Neoplatonist philosopher Iamblichus (245–325 AD) thought in this vein. He describes the different mathematical divisions of the sky that serve as the basis for distinctions in astrological meaning, deriving them top-down from the 'One'.

> Then marking off the heavens into two or four or twelve or thirty-six portions, or twice those numbers, or effecting some other sort of division, they assign to these portions authorities either more or fewer in number, and again they place at their head the One who is over them all. And thus the teaching of the Egyptians about the principles embraces everything from on high right down to the very lowest.[2]

'One' is symbolised in a horoscope diagram by the invisible point at its centre. It represents the place where the life force spirals into physicality, and as this occurs an invisible geometric matrix organises the emerging energy into a twelve-fold pattern. We can imagine the primal energy as pure white light which, as it radiates out into a widening circular beam, becomes refracted into twelve colours each with a different vibration.

The order of the colours is always the same, and in this respect the zodiac is like a rainbow, which always has the same shape when it appears with its seven colours in the same sequence. Plato would say the rainbow in the sky is a manifestation of the eternal Form of the rainbow residing on the plane of Ideas – and that could apply to the zodiac too.

As we have seen, in their purest manifestation archetypes are numbers and as such lie beyond sense perception, though they are accessible to the intellect. Numbers determine the interrelationships of the archetypes in the zodiac on the quantitative level, and also serve as keys to their qualities. Numbers, as we have seen, become emotionally meaningful as musical chords, and when personified as gods and goddesses they become accessible to the imagination. In the same way the numbers of the zodiac become emotionally, imaginatively and sensually meaningful through astrology.

As we have seen, Egyptian metaphysics identified twelve Neters as creators of our space-time universe, each having its own number and musical note. All the world's phenomena could be divided among the twelve categories they governed.[3] An astrologer would see them as the twelve astrological archetypes, symbolised by the twelve signs of the zodiac. (Although it has been argued there should be thirteen signs because the line of the ecliptic runs through thirteen constellations of stars, but this is bottom-up thinking!)

That the number of the astrological archetypes is twelve is supported by the ancient association of the twelve-sided dodecahedron with the twelve cosmic forces at work in our universe (see p. 69), and also by the fact that the number of zodiac signs in astrology the world over has always been twelve, although their animal symbols vary on the different continents. The word 'zodiac', by the way, derives from the Greek word *zodiakos* which means a circus of little animals – although there is nothing little about these mighty archetypal powers!

The timeless order

We can therefore see the zodiac as an image of the timeless order of the twelve astrological archetypes within the circle of eternity. They are represented in it by the signs, each being a thirty-degree section of the 360° circle. The eternal geometric patterning of the archetypes' inter-relationships is illustrated in this matrix (see Figure 11). It depicts an ideal 'alpha-omega' state of perfect balance with nothing in excess and nothing lacking. However,

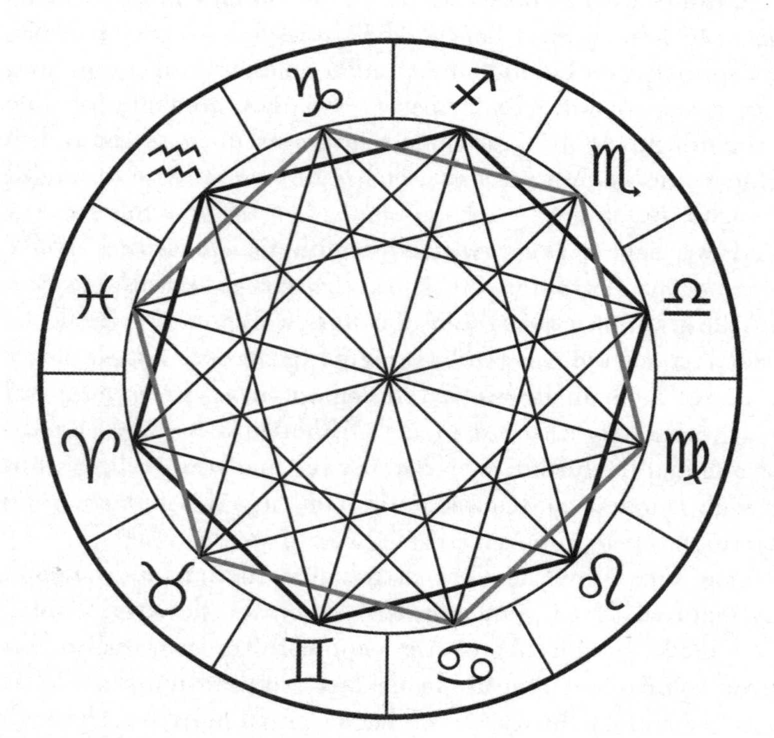

Figure 11. The geometry of the zodiac.

seen from within our space-time universe, where we experience a continuous flow of varying archetypal combinations, the ideal balance of the proportions is lost.

The alpha-omega zodiac is like the allegorical Temple described in the Biblical book of Revelation – holy Jerusalem. We are told that geometry, numerology and astronomy were reconciled in the Temple's design, and the names and numbers of the sacred principles were built into it in accordance with the proportions of geometric symmetry. It was said that every element of creation was represented in the Temple, and that the principles were all in their correct proportions and in accord as defined by the geometric scheme. In other words the temple was an icon of celestial perfection.[4]

The zodiac matrix was recognised as structuring time. A quotation from the Vedas, which are reputed to be many thousands of years old, runs: 'The wheel with twelve spokes rolls round the sky maintaining order',[5] and astrology, which in its origins is just as ancient as the Vedas – or more so – provides access to the qualities symbolised by the twelve spokes. It reveals the world as the product of twelve archetypal forces whose order of unfolding is predictable once their sequence is known. Although the astrological archetypes in their nature are fluid, the laws of geometric proportion that govern them provide stability and consistency. In fact archetypal matrices could be the only stable factor in our shifting, insecure space-time existence!

I concluded that the zodiac should be seen as a model of the cosmos, and that the archetypes symbolised in it are nodes of meaning on the qualitative level of its geometry. As such they are the basis of the ancient ordering system of the correspondences, used by our ancestors as a standard of reference in every department of life. All the phenomena of human experience can be shared out between the zodiac's twelve categories. And in each category a hierarchy of meaning leads down from the most universally applicable ideas to those most fleeting and inconsequential.

The first characteristic of the zodiac matrix is that it exhibits dual symmetry. For example the thirty degree segments of the circle are alternately male and female in their energy qualities, creating a balanced round of positive (male) and negative (female) forces (see Figure 12).

The archetypes also balance each other in pairs of opposite principles, for example Leo and Aquarius are polar opposites and so are Cancer and Capricorn. In this relationship they are like the positive and negative poles of a magnet, linked by the strong current of energy flowing between them, and both attracting and repulsing one another at the same time.

They also serve to define each other. For example when the Aries archetype is expressed in a personality it manifests as qualities of assertiveness and decisiveness. These are in stark contrast to the qualities of its polar opposite the Libran personality that then appears all the more prevaricating and conciliatory.

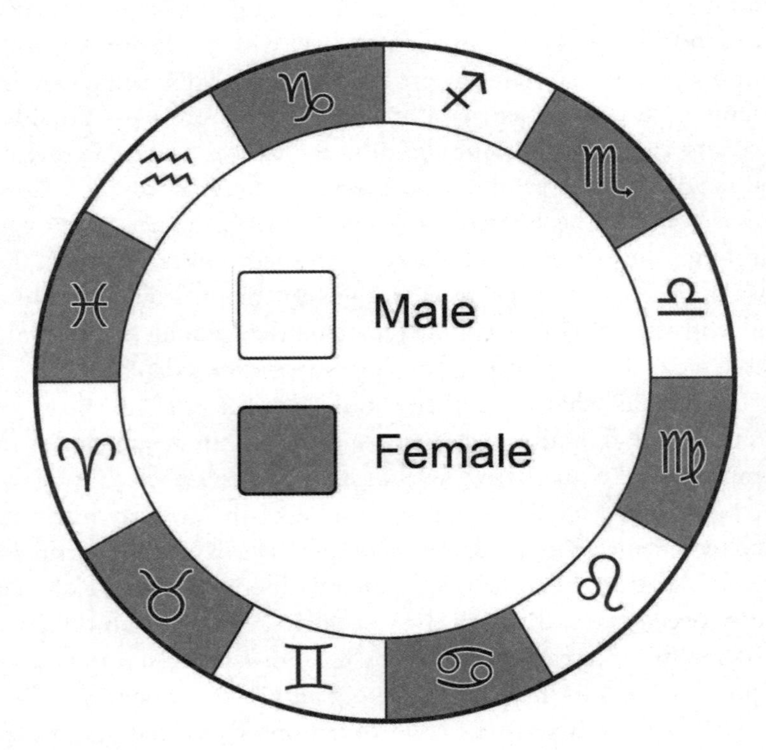

Figure 12. The gender of the signs.

Disharmonious relationships are also created through the zodiac geometry, for example between the six pairs of polarised archetypes and the pairs that lie 90° away from them in the circle (see Figure 13). Cancer and Capricorn are thus at energetic loggerheads with Aries and Libra, and the quality of the right-angled relationships between these four signs is tense and edgy. The same applies to the relationships between Taurus, Leo, Scorpio and Aquarius, and Gemini, Virgo, Sagittarius and Pisces. Altogether there are three large crosses in the zodiac, which we also depict as squares, and which in astrology are called the *cardinal, fixed* and *mutable* grand crosses (see Glossary).

Although the archetypes clash in their square relationships, they harmonise in the triangular relationships created by the zodiac geometry. For example Aries, Leo and Sagittarius form

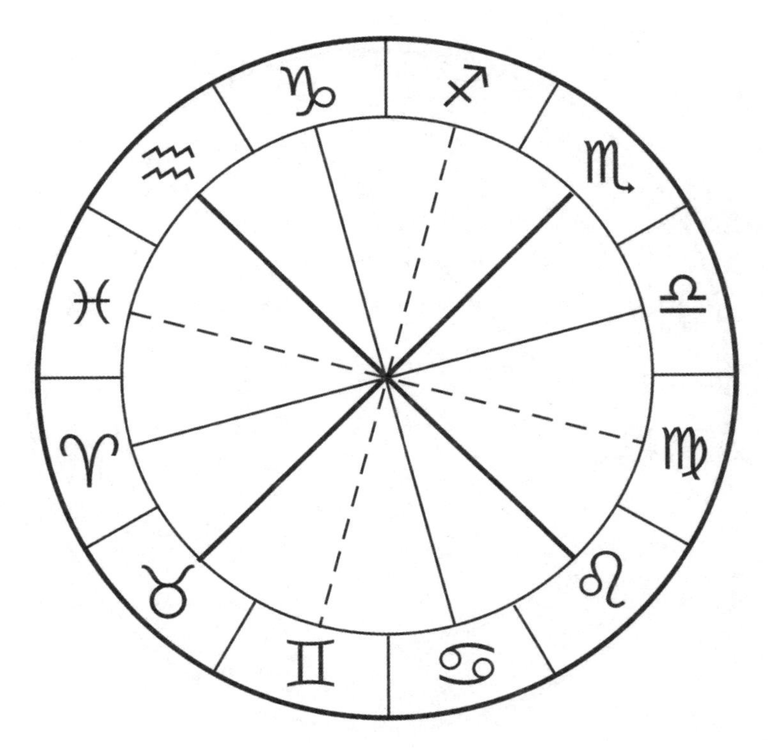

Figure 13. The three grand crosses.

an equilateral triangle within the circle, a figure evocatively described as 'divinity resting within itself'. In triangles the energy flows easily and evenly, and the archetypes related in this way harmonise and support each other. In the zodiac there are four large equilateral triangles which in astrology are called *grand trines* (see Glossary and Figure 14).

The basis of triangular relationships is similarity of element and the basis of square relationships is similarity of *modality* (see Glossary). Added together these geometric relationships produce all the primary qualities. Thus heading the hierarchy of meanings within the archetypal categories are those determined by the gender, element and modality of an archetype – all of which are defined by the positioning of the signs within the zodiac circle.

Figure 14. The four grand trines.

As the word signifies, the twelve signs are 'signs' (meaning 'symbols') of the twelve archetypal principles. Therefore they're not the same thing as archetypes, though I may sometimes use the two words interchangeably when speaking of the zodiac. The main point I'm making is that the geometry of the zodiac is the source of the values and meanings of the signs; it defines their primary qualities. And because these primary qualities are 'objective' in the sense of belonging to the deeper order of the cosmos, each time an interpretation based on them turns out to be true, it's proof that these cosmic values really do exist!

To return to the levels of reality depicted in Figure 1 (p. 22), I'm suggesting that the zodiac is one of the matrices of creation on Level C that pattern what comes into existence on Level A, and also pattern our mental and emotional life on Level B. I'm suggesting too that the astrological archetypes manifest in the human sphere through the medium of the human mind, which we should see as a fractal of the universal mind. I can imagine a mini zodiac matrix implanted in each person's unconscious, directing them from behind the scenes as it were, without their being aware it's happening. However, to those versed in astrology the ordering of the zodiac matrix is recognisable within the maelstrom of particulars in the flux of their life experience.

The belly button of the world

We saw in Chapter 3 how the tetraktys gives a mathematical account of the descent of the archetypes into the material world, and how in mythology this descent is enacted in various ancient creation stories. In the latter abstract numbers are personified as deities, which allows their archetypal qualities to be symbolised in human (or super-human) terms. We have also seen how the geometry of the zodiac leads to the differentiation of the meanings of the twelve astrological archetypes. We will now go on to examine the symbolism of the different components of an astrological chart in the light of the underlying geometric matrix.

At the top of the tetraktys is the One, which corresponds symbolically to the centre point in the chart from which all emerges. It's the spot where astrologers, in the old days before computer print-outs, would place the point of their compasses before drawing a circle around it. The circle expands from the point as in Figure 15, which also happens to be the astrological *glyph* for the sun (see Glossary). This symbol has two components – a centre point and a periphery – which stand respectively for unity and duality, or eternity and the cyclic processes of time.

As the meanings of astrological glyphs can be better understood experientially, I suggest trying the following exercise. Close your eyes and take your consciousness inwards, moving deeper and deeper until you reach what feels like your personal centre. Imagine this as a point of light, shining like a tiny star in a dark night. This is the centre of your being, the source of your life energy, and also the door through which you came into the world. Being a mere point, the light there belongs neither to space nor to time. In other words it's the dimension of you that lives in eternity.

Now imagine the light at your centre expanding like a balloon to become a globe of light that encircles and encompasses you. Its periphery, representing your outer self, marks the limits of your expansion into the world. It's where you end and the world begins. And, with the centre point representing your eternal being, the area of the circle between centre and periphery stands

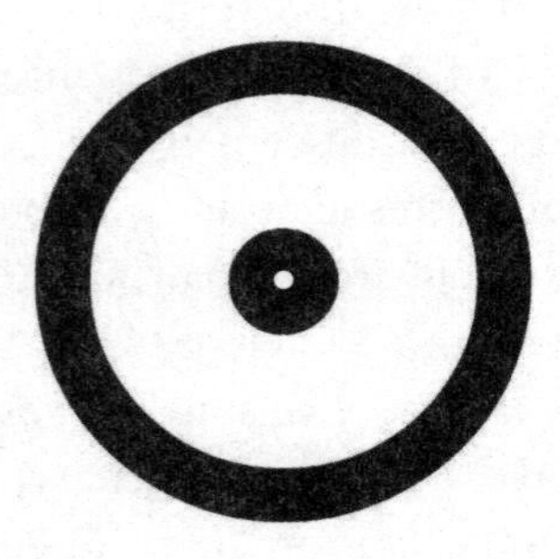

Figure 15. The sun glyph.

for your personality with its conscious and unconscious levels. This area becomes filled up with your life history. Finally, at the end of your life, the circle will shrink back into the point and you will disappear from the space-time world. So allow this to happen now in your imagination, and experience finally disappearing into your centre – into the tiny point of light.

According to cutting-edge cosmology the universe is omnicentric. And as everything has expanded out of one point, so every point within it is, and was, the centre of the universe. However, as soon as we take up a position and see things from a particular perspective, a set of consequent mathematical-geometric relationships kicks in. Incarnation in the three-dimensional world thus naturally incurs the geometry of time and space as a consequence, and this is symbolised by the expansion of the point in the centre of Figure 15 into the full circle.

What we know about ancient Egyptian temple founding ceremonies suggests they consciously mirrored the creation of the universe. For creation to happen in the first place a centre is needed, and this was symbolised for the Egyptians by the 'benben' stone on the sacred mound of Heliopolis. The Greeks also had their 'omphalos' stones, one being the pillar in the centre of Athens dedicated to the 'twelve gods'. (In this connection I recalled that in Plato's ideal state the land was divided into twelve parts by twelve radial lines leading out in twelve directions from such a centre creating the zodiac matrix in the surrounding landscape.)[6]

An omphalos stone can still be seen today at the temple of Delphi in Greece, described by the ancients as 'the belly button of the world'. And I suggest that you see the centre of your birth chart as your personal omphalos, symbolising the place where the fount of your life energy wells up from a deeper source.

To initiate creation on the physical plane a centre is needed as a fixed focal point around which a geometric 'field' can extend. Physicists today are investigating the Higgs' boson which supports the accretion of particles to produce mass. It appears that form, in the sense of abstract geometric patterning, could always arise together with substance, and that material creation occurs through densification. The energy needed to fuel this process comes from the heat produced by velocity. And that explains why the planets in our solar system revolve round the sun at such incredible speeds, and why the sun revolves round the focal point of the galactic centre at an unimaginably high velocity.

So the point in the centre of your chart stands for your internal energy source. However you have a second source of life energy that is symbolised by your *Ascendant* (see Glossary) on your chart's eastern horizon (see Figure 16). Your Ascendant is where the life force enters you from the outside – in other words from your environment. And these two sources of energy, your centre point, which is also symbolised by the sun in your chart, and your Ascendant together contribute in a large part to the way you are and to how you choose to live.

Dividing the light from the darkness

The second step in the numerical evolution symbolised by the tetraktys occurs when the One becomes Two. The geometer takes this step when he marks a second point at a distance from the first, and draws a line between them. In your chart this line is the central horizontal line that, passing through the centre point, links the Ascendant with the Descendant on the opposite side of the circle (see Figure 16).

One is an odd number whereas Two is even; and, according to Pythagorean number theory, odd numbers are male and even are female. So dividing the One to make Two produces the schism at the root of all further dualities, which is the separation of male and female energies. An atom decays, I have heard, into two particles which move off in opposite directions, and which have opposite spins. They remain however 'entangled' (in contact on an information level), and, when the spin of one is determined, the other necessarily adopts the opposite spin state. Interesting! Because this fact is relevant to the way the opposite signs in the zodiac define each other.

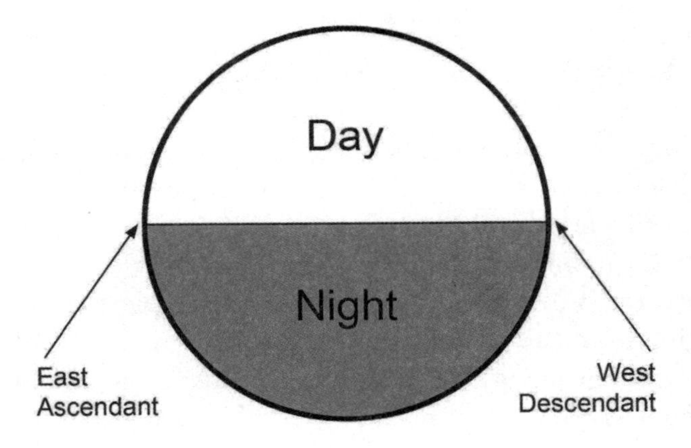

Figure 16. The day and night hemispheres.

Within our time-space world the creative cosmic forces manifest in oppositions of positive and negative energies, polarising like the positive and negative poles of an electric battery. The primary state of polarity is represented in an astrological chart, in the first place, by the two hemispheres – that above and that below the horizon. The basic quality of the upper hemisphere is male, positive and extravert, while the basic quality of the lower is female, negative and introvert.

The sun passes through the upper hemisphere during the day and the lower during the night. Therefore the day hemisphere stands symbolically for our waking state when we are active out in the world, while the night is where we retire into our private space, and enter the inner self when we sleep. This fundamental duality extends in astrology beyond the hemispheres to cover the gender distinctions between the positive and negative signs, houses and planets.

The sun and moon in their roles of lord of the day and queen of the night are central to the basic gender division in astrological symbolism. However over the centuries, and in different cultural contexts, their genders have been swapped round. For example, the moon was seen as male in Egypt and Babylon, and female by the Greeks and Romans. That this can happen points to the gender assignations used in astrology being more flexible than is supposed; everything including astrology it seems is in flux.

Perhaps the explanation lies in the fact that in both Hindu and Egyptian theology the main deities are hermaphrodites having both a male and a female form. Similarly in traditional astrology the five planets (excluding sun and moon) have both male (positive) and female (negative) modes of expression. For example, Jupiter as ruler of Sagittarius manifests positively, and negatively as ruler of Pisces. The important point here, however, is gender symmetry. The sun and moon, representing male and female, are perfectly balanced, being equal in size when seen from the earth.

In modern astrology the sun, now known to consist of burning gases, is seen as the quintessence of male energy, as hydrogen burns continuously within it producing light and heat. In contrast the moon is an apt symbol of female energy as it's solid like element earth, and it affects the water on our planet through its gravitational pull. The basic distinction between the two as I see it is that the sun radiates light whereas the moon absorbs and reflects it. And this is an indication of how the male and female energies can be distinguished in astrology.

Figure 17 shows the basic qualities arising from the gender distinction of the archetypes as complementary fields of meaning.

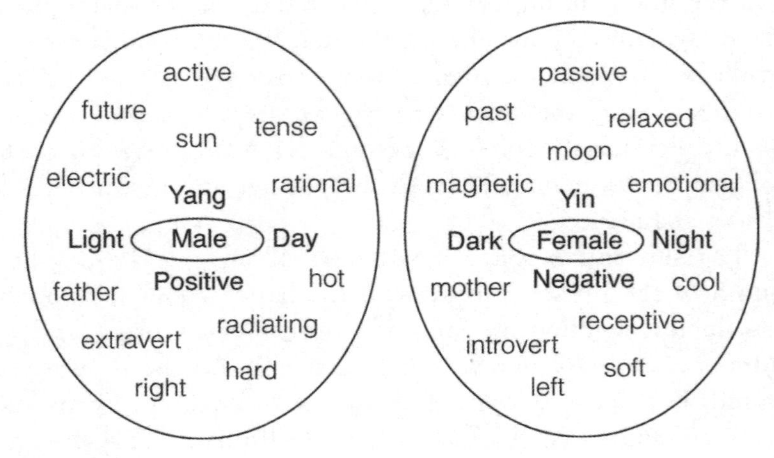

Figure 17. Gender fields.

The words in the two fields parallel each other as opposites, and also serve to define each other. And mutual definition is basic to the differentiation of meaning in astrology, as is shown in the geometric relationships between the twelve principles in the zodiac. For example male energy is only hard in contrast to female energy which is soft, and a left-brained type of mind is only rational in contrast to a right-brained mind which is intuitive.

Your birth chart will indicate which of the two energy forms – male or female – dominate in your make-up, and it will also show how far they are out of balance. Counting the number of planets above and below the horizon is a start, but then you should go on to check whether more female or more male signs are highlighted through being occupied by planets. Later you would also bring in the gender of the occupied *houses* (see Glossary) and the gender of the more dominant planets in your chart to get a fuller picture. Balancing and harmonising the female and male energies physically and psychologically is self-development work and so central to the agenda of those on a spiritual path.

The qualitative distinction between the two hemispheres in your chart leads to a different interpretation of your planets depending on whether they are above or below the horizon. It's also important to bear these gender distinctions in mind when we later interpret *opposition* aspects (see Glossary), which, unless they are fused with the horizon line, will always link one planet in the upper with one in the lower hemisphere. The male-female polarity of the hemispheres accounts for the dynamic of combined mutual attraction and repulsion that characterises this type of aspect.

Self and Other

To return to the comparison between the ground plans of ancient temples and astrological chart diagrams, when a henge was being constructed in Neolithic times the first step taken was to divide the site by an east-west line. The surveyors of those days achieved this by using poles as sighting sticks to determine the angles of sunrise and sunset by measuring the shadows they cast.

They were aware that, at the equinoxes, the shadow of the rising sun cast by one stick would perfectly align with the shadow of the setting sun cast later that day by the other. And this line would be marked out on the ground. An original east-west alignment has been found at Stonehenge marked by post holes that archaeologists believe could date back to around 8000 BC!

The creation of a henge was an act that brought sacred geometry down to earth by translating it into a figure on the ground that was coordinated with a certain latitude. And the ancient Egyptian hieroglyph for latitude is two sphynx-like lions, representing the eastern and western horizons, with the sun rising between them (see Figure 18). They also symbolise the Neters Shu and Tefnut, who in the Heliopolitan creation story were the first of the coupled principles to arise from the totality of Atum and to manifest the number Two.

In their role of 'guardians of yesterday and tomorrow' the lions of the horizons protect the consistency of the daily passage of the sun from East to West.[7] Thus there is also a duality between the left side and the right side of a chart diagram – left with the

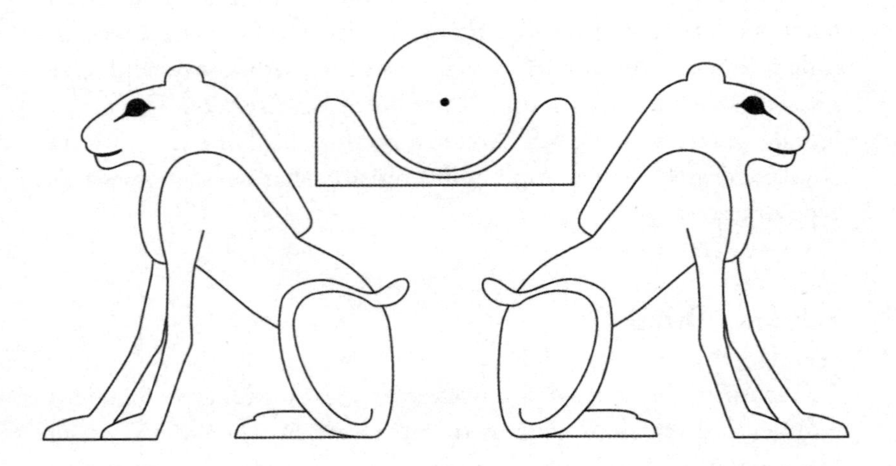

Figure 18. The latitude hieroglyph.

Ascendant representing East, and right with the Descendant representing West, its polar opposite.

So why is the Ascendant in a chart so important? The horizon that divides earth and sky is also where earth and sky meet, a fact of great spiritual significance in ancient cultures. In Egypt and Babylon the heliacal risings of significant stars were not only observed to determine the passage of the seasons. A star rising in the East was seen as being reborn out of the underworld. And I suggest that this focus on the eastern horizon by ancient astrologers as the place of the emergence of new, vibrant life led to the prominence given to it in astrology. The degree of our Ascendant sign symbolises a main energy source as we have seen.

Also relevant here is the fact that, just as sunrise is the beginning of the day, so the sunrise point in our chart symbolises the beginning of our life's day. And, according to ancient wisdom, in the moment when any enterprise is begun its whole future course is contained. This piece of wisdom lies behind a more advanced astrological technique that uses the measure of one day equals one year for prediction. Thus our Ascendant degree is like the acorn out of which the oak tree of our life has grown. And similarly our natal chart, cast for the moment of our first breath, will contain our life in a nutshell.

In astrology the Ascendant is seen as the place of the Self, and the Descendant, which lies opposite it, is the place of the Other. Self and Other are separated by the horizon line which puts 180° between them. It therefore follows that usually the qualities symbolised by our Ascendant are traits we can easily identify with; whereas the qualities symbolised by our Descendant are not. We confront them instead in our opposite numbers in life.

If the Ascendant is interpreted as the birth place of the ego-self – the part that's convinced it's a separate individual – our Descendant indicates a way to transcend this illusion of separation. Its symbolism can direct us towards a path of love whereby self-serving is transformed into service to others, ego-consciousness is overcome and Self merges ultimately with Other in mystic union.

CHAPTER 6

Squaring the Circle

A Stone Age calendar machine

An interesting ancient astronomical work, *The Book of the Courses of the Heavenly Luminaries,* which forms part of the apocryphal *Book of Enoch,* has survived from classical times. Christopher Knight and Robert Lomas refer to it in their book *Uriel's Machine,* in which they investigate the origins of astronomy.[1] This ancient work, possibly deriving from the second century BC, though its contents could be far older, recounts how Enoch was instructed in astronomy by the archangel Uriel in a faraway land to the north. And Knight and Lomas make a case for it being the British Isles. The following quotation is relevant to the way the zodiac should be understood:

> In like manner twelve doors Uriel showed me, open in the circumference of the sun's chariot in the heaven, through which the rays of the sun break forth: and from them is warmth diffused over the earth, when they are opened at their appointed seasons.[2]

Here the sun is seen as a round chariot, rather like a flying saucer, with twelve doors in its circumference through which its rays shine forth in turn in the course of a year. Thus the sun, standing for the One, has twelve different facets. As it beams through each door in turn its light is modified by different characteristics. And the opening and closing of the doors explains the rhythm of the sun passing from one zodiac sign to the next with the passing months. Thus the zodiac circle becomes an extension of the being of the sun, and its sections express twelve different facets of the sun's nature.

Enoch is also shown portals through which the sun appears at various times of the day and year, which Knight and Lomas suggest could be the gaps between the stones of a stone circle such as at Stonehenge or Avebury. They envisage the Neolithic astronomers using these gaps, that were mathematically calculated when the stones were positioned and aligned, for observing and calculating the passages of sun, moon and stars.

> And I saw six portals in which the sun rises, and six portals in which the sun sets. Six in the east and six in the west, and all following each other in accurately corresponding order, also many windows to the right and left of these portals.[3]

If the 360° degree circle is divided into twelve sections, and each section is further divided into three, thirty-six windows could result – the number of the astrological *decans* (see Glossary). The Egyptian sky was divided into thirty-six decans that are still in use in astrology today, and each was marked by a bright star. In addition to their functional time-keeping purpose these divisions had astrological significance, as each decan was associated with a ruling decan spirit that had particular traits. In other words the Egyptian decans were subdivisions of the signs, and their ruling spirits were facets of the twelve astrological archetypes.

Enoch describes how during the year the sun was seen to rise within an area that was six windows wide along the eastern horizon, and similarly was seen to set through a series of six windows on the western side. Twenty-four windows would then remain to complete the circle of thirty-six decans, and these could be the 'many windows' that Enoch saw to the right and left of these two sets of portals. Knight and Lomas believe this text derives from an account of how Neolithic stone circles were used as 'calendar machines.' What is important for us, however, is the mathematical 'genealogy' of the astrological principles implied here.

The passage from Iamblichus quoted on page 94 creates a link between such ancient astronomical practices and the astrological chart. A circle can be divided into two as we have seen, producing the two hemispheres of day and night. It can also be divided into four to give the four directions and the four cardinal principles (see Figure 19), or into twelve as in our present zodiac, or into thirty-six as in the more finely differentiated ancient Egyptian map of the sky. In this chapter we are going to investigate the fourfold division.

Although modern western charts consist of two concentric circles, the outer being the wheel of the signs, and the inner the wheel of the houses, in the past their shape was square, and in India they are still square today. Both circle and square are symbolically meaningful – the circle symbolising eternity, and the square generally seen as standing for our space-time world.

All circular chart diagrams also contain squares created by the four *angles* (see Glossary) as their main structure. For a geometer the task of 'squaring the circle' has always been a testing mathematical challenge, as it involves combining square and circle, whose perimeters must be exactly equal, in one scheme of proportion. However we know Stone Age architects were capable of this feat, as squared circles have been identified in the ground plans of many ancient sacred sites.[4] Perhaps it was the very inclusion of this geometric figure that made the sites sacred, as the act of squaring the circle always carried a mystical significance.

As we have seen, 'temples' such as Stonehenge and Avebury were built according to proportions laid down in what Michell calls 'the lost canon of the ancient world'. Their ground plans, which include the figure of the squared circle, reflect the primordial vision of an eternal cosmic order.[5] Michell draws our attention to the six two-way stone alignments at Stonehenge to the eight extreme positions of the moon and the four of the sun, making twelve positions in all.[6]

In the context of a geometrically-based cosmological thinking, the belief may have been that uniting the figures of the circle and the square brought heaven down to earth. And this act could have also symbolised reconciling body and spirit. It was logical to see the circle as representing spirit, because the measure of its perimeter, which is *pi* times its diameter, defies definition on account of the irrational nature of *pi.* In contrast, the perimeter of a square is precisely four times the length of one of its sides. Thus it became the symbol of material order, while when a circle was drawn round it, neatly uniting the polar opposites of its corners, this could have symbolised the transcendence of duality for mystically minded geometers.

The geometry of space and time

Ancient temples were also planned in relation to their surrounding landscapes, whereby alignments to the four cardinal directions were of vital importance. The four corners of the Great Pyramid are aligned north, south, east and west with astounding accuracy, and cardinal alignments have been found in all the Egyptian pyramids so far surveyed. How did they do it? Well, as we have seen, east and west could be determined at the equinoxes using the sunrise and sunset points. And north and south were also determined astronomically through charting the sun's path during the day and the passage of the stars at night. The sun's zenith always lies due south, as Neolithic astronomers knew, and north could be found by locating a point on the horizon immediately below the celestial North Pole around which the circumpolar stars revolve.

Once established, the north and south directions were joined to create the vertical line of a central cross that divided a temple site into four quarters. It's interesting that the Egyptian hieroglyph for city is a circle with a cross in it, which probably reflects the geometry of their town planning. And ancient cities can still be found today – Beijing and Cusco are examples – with four gates in their outer walls, each facing in a cardinal direction and connected by two main roads that form a cross.

These are examples of crosses created on the earth, but the ancient astronomers also identified a great cross spanning the sky that is formed by the four bright stars Aldebaran, Antares, Formelhaut and Regulus. And, because the sky appears to revolve around us once in twenty-four hours due to the earth turning on its axis, every six hours on an average one of these four bright stars will rise on the eastern horizon – for example at sunrise, midday, sunset and midnight – the peak moments in the twenty-four-hour cycle of the sun.

This cycle was understood as, what would be called today, a 'fractal' of the sun's annual cycle of twelve months. Thus the four peak times of the day paralleled the four peak times of the year – the two equinoxes and the two solstices. Figure 19 shows these correspondences between the angles in a chart, the cardinal directions and the cardinal times of the day and year.

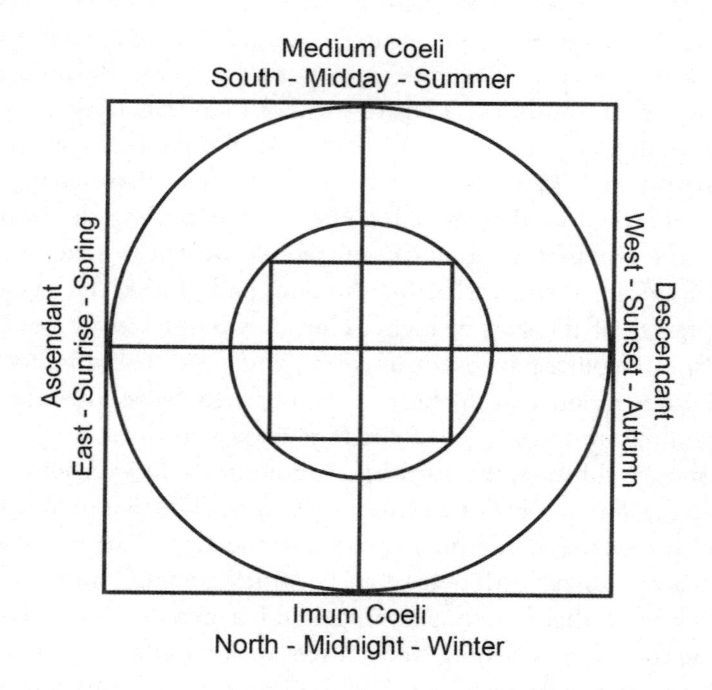

Figure 19. The square of time and space.

In the ancient cultures each direction had its deity and its symbolic significance, which must have contributed to the meanings we ascribe in astrology to the four angles in a chart. North, for example, symbolised immortality, as that's where the circumpolar stars are found – the 'imperishable' stars that never rise or set. And as the north was also associated with the midnight hour in the sun's diurnal cycle, and the darkest time of the year of the winter solstice, it came to symbolise quiescence. In contrast the south as the place of maximum light, where the sun reaches its midday zenith, and its highest point in the sky at the summer solstice, symbolised fruition.

The solstices and equinoxes, which were considered times of great efficacy, were celebrated in Neolithic temples with elaborate feasts.[7] The orientation of the Knowth passage-tomb in Ireland, for example, points to festivals being held on this site that marked the spring and autumn equinoxes, whereas the alignment of the nearby Newgrange passage-tomb to sunrise at the winter solstice highlights that time of the year.[8] Both structures are located in a Neolithic 'sacred landscape' situated within a bend of the river Boyne.

The sacred geometry of an astrological chart is based on squared circles and encircled squares as shown in Figure 19, where the arms of the central cross correspond to the cardinal directions and to the chart's angles. They cut the outer square at the four places where it touches the outer circle, and perhaps these symbolised contacts between heaven and earth for the ancients. When the sun on its annual round arrived at one of these points, corresponding alignments built into an ancient temple would be 'activated'. Then the rays of the sun would shine down the dark passageways into the sanctuaries creating celestial pathways of light.[9]

The Delphic oracle was only active once a year in earlier times, when, according to legend, streams of a 'magnetic' current would flow accompanied by manifestations of spirit. Later on this became a more frequent occurrence (perhaps for commercial reasons). But we assume that originally sacred sites in general were believed to have specific times when they were 'live' in this sense. And it was the task of the astrologer-priests to calculate the peak moments when they would be 'plugged in' and the current would flow, so advantage could be taken of the beneficial energies then available.

As we saw in Chapter 2, in Egyptian mythology (that was possibly inherited from the same civilisation that built the megaliths of Atlantic Europe) sky and earth were represented as the lovers Nut and Geb, who had been tragically torn apart and were yearning to reunite. And it may have been at these peak moments – geometrically represented by the intersection points between square and circle – that Nut and Geb metaphorically mated, and an energy exchange between earth and sky was triggered.

In our two-wheeled horoscope the realm of Nut (the sky), is represented by the outer wheel, and the realm of Geb (the earth) by the inner wheel. And these ideas and associations, suggested by what we know of Neolithic astronomy, help to explain why the angles – Ascendant, Descendant, *MC* and *IC* (see Glossary) – in a chart represent such energetically loaded points.

If we join up the four corners of the two squares in Figure 19 diagonally, a saltire cross is formed (see Figure 20). Its corners mark the northeast, northwest, southeast and southwest, and correspond in the calendar to the cross quarter days that fall during the first weeks of February, May, August and November. They were (and in some places still are) celebrated with the Celtic festivals of Imbolc, Beltain, Lusaghnadh and Samhain respectively. On the other hand the points where the outer square and circle meet mark the main Christian festivals.

The times of the Celtic feasts associated with the saltire cross are astronomically significant as they mark the days of the year when the speed of the movement of the sunrise point along the eastern horizon changes. After Imbolc and Lugnasadh this movement gathers speed, and after Samhain and Beltain it slows down. The Stone Age astronomers, who would have had these points plotted along their local horizons, certainly knew about this phenomenon, which no doubt had religious significance for them.

Imbolc, Lugnasadh, Samhain and Beltain are celebrated during the months of Aquarius, Leo, Scorpio and Taurus respectively – the four fixed signs. While the Christian festivals of Christmas,

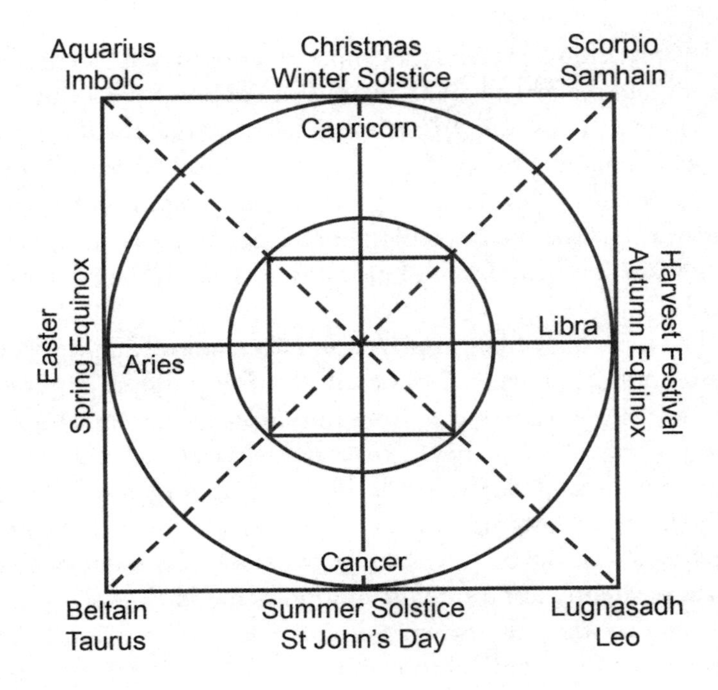

Figure 20. The cycle of pagan and Christian festivals.

Easter, John the Baptist and Harvest Festival are celebrated during the months of Capricorn, Aries, Cancer and Libra respectively – the four cardinal signs. So can we relate the religious ideas and symbolism inspiring these festivals to the eight astrological archetypes manifesting at these times of year? Probably yes!

The gods of the directions

The astrological archetypes that manifest as lords of the four directions have been anthropomorphised down the ages as different gods and goddesses. For example in Egypt they became 'the four sons of Horus', 'guardians of the celestial quarters'. These deities would be invoked during the founding ceremonies for new temples and summoned to each of the respective corners.

An inscription at Dendera temple describes such a ceremony.[10] We are told that the Pharaoh himself would break the ground and bury four amulets, one in each quarter, to represent the sons of Horus.[11] This was believed to cause the creative forces of north, south, east and west to emerge from the ground and together form the spiritual body of the temple. At the same time the act symbolised the structuring of primordial chaos at the birth of the world.

The four sons of Horus played an important role in the process of mummification when they became the four canopic jars used to contain the viscera extracted from the corpse before embalmment. This procedure was believed to ensure that Horus, through his sons, would renew the power of the four main organs of the body of the deceased – stomach, liver, intestines and lungs. (The heart, which was seen by the Egyptians as the seat of consciousness, was always reverently left inside the mummified body).

It's likely that all ancient cultures had their versions of the rulers of the four cardinal directions. In India, for example, these were the gods Indra (East), Varuna (West) Kubera (North) and Yama (South). Their unification within the figure of Brahma, the One, is symbolised by representations of this creator God with four heads each facing in a different direction. The personality traits and lore attached to the different deities of the directions must derive from the fields of meaning associated with the four cardinal cosmic archetypes. And we can identify Indra, for example, who in Hindu mythology is the god of war and rules the east, as a personification of the Aries astrological archetype.

The builders of the British Neolithic temples would have inherited their sense of the qualitative differences of the four directions from their hunter-gatherer forbears. These were the nomadic tribes who roamed the north European plains at the end of the last ice age, and whose sense of direction, sharpened by survival needs, must have been well-defined. Even in today's world, where most of us have lost the sensitivity to natural forces that our ancestors took for granted, we can be subtly affected by the different energies of the four directions. For example some people find they can only sleep with their head in the north and their feet pointing south!

The East where the sun rises has always been associated with birth (or rebirth), and seen as especially significant for this reason. Hindu temples and Christian churches are built facing east, and Greek temples were aligned along an east-west axis. In contrast the west where the sun sets has from time immemorial been associated with death. Thus at Thebes (present-day Luxor) the metropolis of the living was built on the east bank of the Nile, while the necropolis of the dead was constructed on the opposite west bank.

In many cultures the north was viewed as a source of unfavourable influences, with north-south routes across the countryside having the spooky reputation of being 'ghost paths'. There was also the custom in the Middle Ages of bricking up the north doors of Christian churches in order to prevent evil spirits from entering. And the Egyptian god Set (or Seth), ruler of the north, is a forbear of Satan as his name suggests. In contrast to this gloom the south is all sweetness and light. For example the *Book of the Heavenly Luminaries* tells us that through the portal of the south, 'there come forth fragrant smells, and dew and rain, and prosperity and health'.[12]

In ancient China, where the city gates were given evocative names, the northeast gate was called the gate of ghosts and ancestors. In this context it's interesting to note that the main entrance to the central stones on the Stonehenge site is from the north-east, which confirms the view held by contemporary archaeologists that the site was used to honour the dead with festivals involving ancestor worship.

Fire, water, air and earth

Two plus Two is Four, and Two times Two is Four. Two becomes Four in the mathematical generation of the archetypes by the male archetypal energy differentiating into two types symbolised by the elements fire and air, and the female archetypal energy into the types symbolised by water and earth. The four elements, basic symbols in astrology, describe differences on the qualitative level of the geometry of the four equilateral triangles contained in the zodiac matrix. Below is a table showing how the astrological archetypes divide into these categories.

<table>
<tr><td rowspan="2">Male Energy</td><td>Fire</td><td>Aries</td><td>Leo</td><td>Sagittarius</td></tr>
<tr><td>Air</td><td>Gemini</td><td>Libra</td><td>Aquarius</td></tr>
<tr><td rowspan="2">Female Energy</td><td>Water</td><td>Cancer</td><td>Scorpio</td><td>Pisces</td></tr>
<tr><td>Earth</td><td>Taurus</td><td>Virgo</td><td>Capricorn</td></tr>
</table>

So what about the fifth element? you might say – the Chinese have five elements, and Plato also believed there were five. He claimed that five polyhedra were the geometric building blocks of matter (see p. 68), and, when he described the qualitative level of their geometry by assigning each to an element, he paired the dodecahedron with ether. Ether, however, as its name suggests is ethereal and belongs to a higher dimension, whereas the other four that manifest on the border line between body and soul, are both physical realities and psychological symbols.

Seen psychically, the life force spirals into our universe from an invisible dimension, and on entering densifies into four primary states. These are fire producing light and heat, air producing the gases, water producing the liquids and earth producing everything solid. And out of these four primary states all else emerges. Because the macrocosm and the microcosm mirror each other – which, you'll remember, was the big secret that the axiom 'as above, so below' revealed – our experiences of the elements in the outer world of Level A (p. 22) can be used as symbols and analogies for interior experiences and psychological states on Level B.

We are composed of a system of bodies interfacing one another, of which the physical body is the one that's visible. The rest are subtle. Each of our bodies has a different level of vibration, and is surrounded by an information field. The information contained in these fields is the source of our habitual patterns of acting, feeling and thinking.

Our physical body can be seen as our personal earth, our emotional body our water, our mental body our air and our spiritual body our fire. The four elements that are found in the macrocosm of Nature are paralleled in the human microcosm by the four 'humours', which in medieval psychology were seen as the source of the four temperaments – the choleric (fire), sanguine (air), melancholic (water) and phlegmatic (earth).

Our fire body is the seat of the patterns that influence our desires and intentions, and it's also the source of the energy that enables us to achieve things. Our water body is the seat of our imagination, emotions and intuition, enabling us to form emotional bonds with others. Our air body is the root of our mental faculties allowing us to learn, think and communicate, while our earth body is our material grounding, and provides

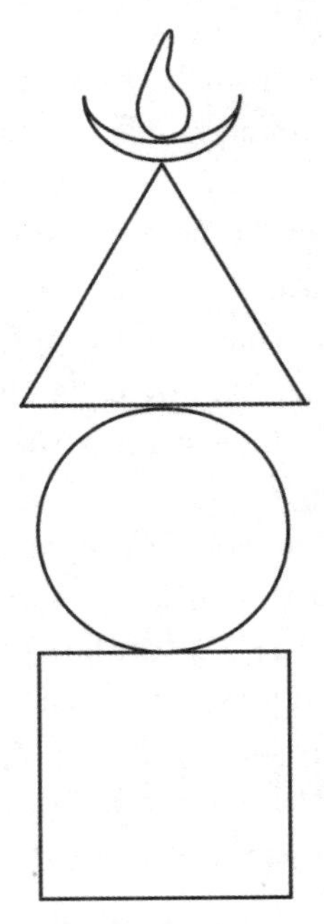

Figure 21. The geometry of the Buddhist stupa.

us with the practical skills needed to manage our material environment.

In the Buddhist stupa the hierarchy of the elements and the bodies is symbolised, rising from the most dense at the base to the most refined at the apex (see Figure 21). The structure rests on a cube representing earth, on which stands a globe symbolising water. The element air is represented by the tetrahedron resting on the globe, while the flame burning in the bowl at the apex is fire. The stupa symbolises the alchemical transformation process of those on a path of spiritual development, who are perfecting the 'great work' of inner alchemy by becoming masters of all four bodies.

The four elements, which are powerful archetypes within the cosmic psyche, can emerge dramatically in the visions seen by the Biblical prophets, for example Ezekiel's vision of the four living creatures:

> They four had the face of a man, and the face of a lion on the right side: and they four had the face of an ox on the left side; they four also had the face of an eagle.[13]

Astrologers will recognise the fixed signs of the zodiac in this description, Aquarius, Leo, Taurus and Scorpio respectively, whereby the eagle is an alternative symbol for Scorpio. In Christian iconography the four fixed astrological archetypes are traditionally taken to represent the Evangelists, Matthew (angel), Mark (lion), Luke (bull) and John (eagle).

The four elements that are combined in Ezekiel's vision are also merged in the ancient symbol of the winged sphinx, which has the body of a lion (fire), a human head (air), a bull's neck (earth) and an eagle's wings (water). Seen esoterically the winged sphinx, like Ezekiel's vision of the living creatures, represents the mathematical reduction of the Four into One, and thus symbolises the mystical translation of material creation back into unified light.

As we saw in Chapter 5, there are four equilateral triangles in the zodiac that are formed by the signs that share the same element (p. 100). These four three-sided figures express a combination of the numbers Three and Four. And the origins of the meanings attributed to the four elements in astrology can be traced to these numbers and the geometry resulting from them. Below is a table showing the archetypes divided into the categories that in astrology are called the *triplicities.*

In my experience the most effective way of communicating

Element	Signs		
Fire	Aries	Leo	Sagittarius
Air	Gemini	Libra	Aquarius
Water	Cancer	Scorpio	Pisces
Earth	Taurus	Virgo	Capricorn

an archetype's qualities is to illustrate them as a field of keywords surrounding a central core meaning. A graphic representation speaks to our intuition and imagination as well as to our reason, adding a multi-dimensional richness. So try moving into a deeper understanding of the water archetypes by meditating on the word field below (see Figure 22). See the field as a whole, and allow the words it contains to fall into your imagination like pebbles, and trigger spreading ripples of wider feelings and associations.

Water was seen by the pre-Socratic philosopher Thales as the primary substance from which all creation arose, and it's also the first element mentioned in Genesis, where we're told that God created the firmament by dividing the waters below the earth from the waters above it. These were the primordial waters of Nun in Egyptian cosmogony.

In order to understand the water archetypes try the following exercise. Close your eyes and visualise rain ... Little streams of water are running down hill sides to create lakes that swell into rivers and finally pour into the ocean. Imagine you are swimming in a crystal clear sea ... or taking a cooling shower. Feel the sensation of water on your skin. Now imagine yourself descending

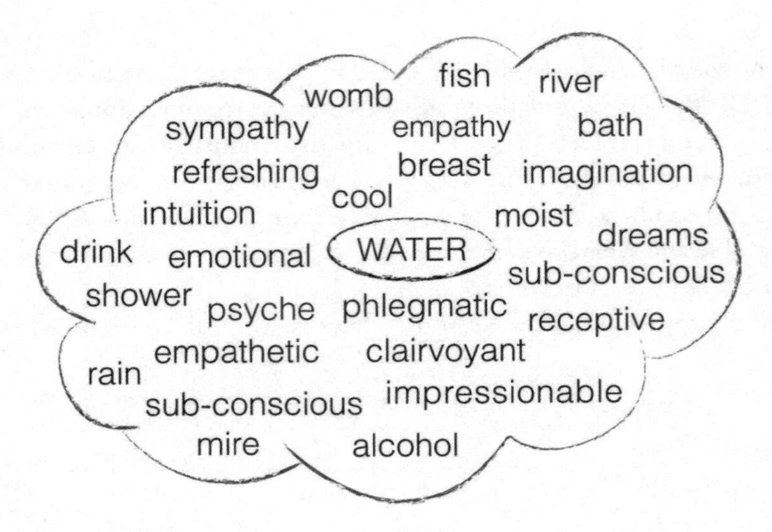

Figure 22. The water field.

into the ocean's depths through shoals of fish – are you afraid of meeting a giant octopus or sea-snake? And finally you find yourself storm-tossed in a tiny boat, rocked by large waves … So what would you say are the characteristics of water?

All your personal associations will be relevant to the meaning of the symbol water. However the ideas you come up with can be distinguished into three types: some will be more narrowly subjective, some will reflect your share in the common human experience of water, and some may indicate the qualities of water as an archetype *per se.* The keywords in the word fields in this book are roughly divided into these three categories.

Although in the context of archetypal cosmology the archetypes have autonomous meanings which are 'out there' in the universe, we co-create an archetype's meaning in the act of interpreting it. In so doing we add the inflexion of our personal experiences together with memories from the bank of human collective experience that we automatically access. Thus an interpretation is a composite of these factors.

To return to water, as it's a female element the natural direction of its movement is downwards and inwards, which, when translated into psychological terms, suggests that a water person (someone with

Figure 23. The fire field.

a water emphasis in his chart) will be more of an introvert than an extravert. Water is passive in the sense that it's moulded by outside forces – it will take the shape of any container it's poured into – thus water people are impressionable and react sensitively to their environments. Water also seeps into things, which symbolises on the inner level the water person's empathy. She will be able to intuitively understand others on a feeling level and sense where they're at.

Another characteristic of water is its power to erode, which could correspond to the emotional pressures that water people often exert on others. A slow dripping wears away a stone! Also, although water is transparent near the surface, its deeper levels are opaque. And its dark depths become an analogy for the subconscious levels of the psyche, where repressed passions and drives can lurk like hidden currents, and erupt as irrational behaviour and violent mood swings.

In order to move more deeply into the meaning of the fire archetypes, close your eyes and imagine a candle flame … a cheerful, crackling fire in a hearth … a bonfire … fireworks … forked lightning and lastly a huge conflagration, a forest fire consuming vast tracts of land. So what are the qualities of fire? (See Figure 23.) Well, it radiates heat and light, which translates into the strong

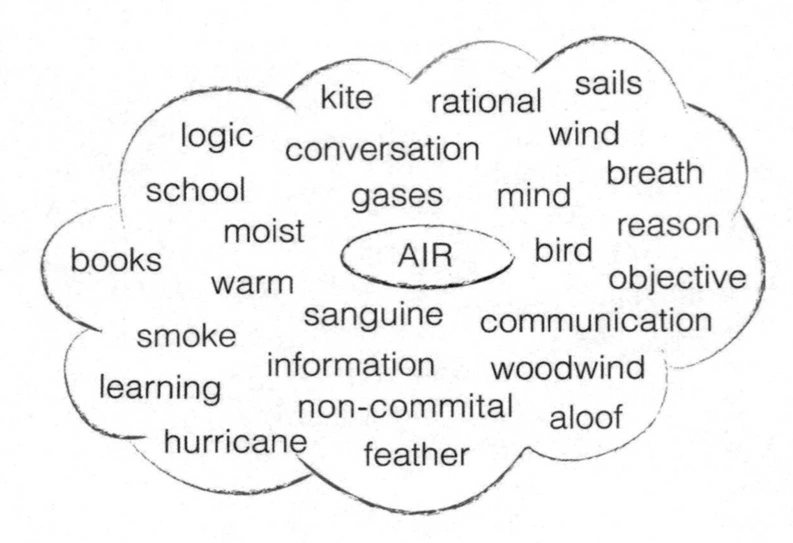

Figure 24. The air field.

vitality and buoyant enthusiasm of the person with a fire emphasis in his chart. He has the power to inspire people, and others will be magnetically attracted to his sparkle and warmth.

Just as fire is expansive, striving ever higher and attempting to reach beyond, so the fire person is difficult to satisfy, desiring to continually achieve more. He craves adventure and is ready to take risks. However fire is also dangerous; it can burn and destroy. Psychologically its heat correlates with anger, which can flare up in the fire person especially if he's thwarted. Also, as fire is voracious, and indiscriminately consumes everything in its path, so the fire person, hot-foot in pursuit of his self-serving ends, may trample on others and leave a trail of devastation in his wake.

To move more deeply into the meaning of the air archetypes, imagine billowing clouds ... a smoking chimney ... wafting autumn leaves. Blow out a candle ... blow up a balloon ... and feel the sensation of a cool breeze on your skin. Finally imagine walking in open countryside when it's blowing a gale and you're being buffeted by storm-force winds ... Then what are air's qualities? (See Figure 24.)

Air, like fire, is a male element and so its natural movement is upwards and outwards. Thus the attention of the air person will

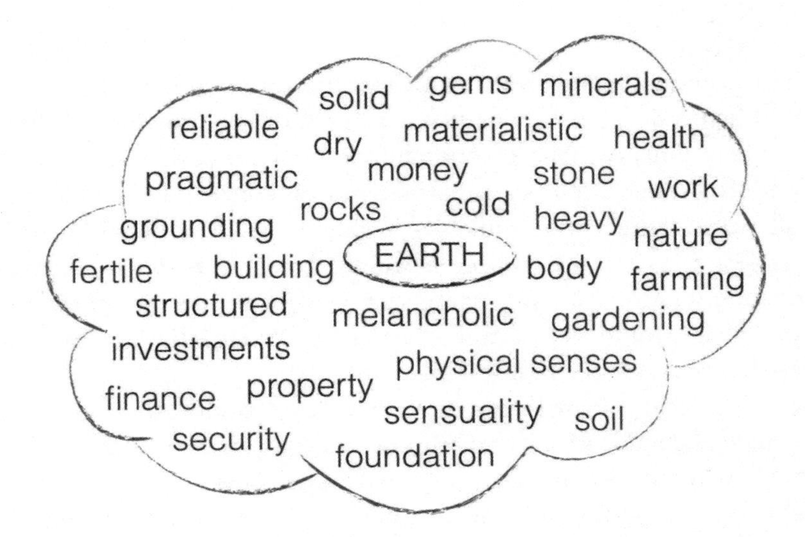

Figure 25. The earth field.

be focused on the outside world and on other people. And, just as the fire person will continually seek new experiences, so the air-person will be on the look-out for new pieces of information and interesting social contacts.

Air is transparent and mobile and it's also ubiquitous; it fills the spaces between things, and forms a layer above them. On the psychological level this correlates with the air person being difficult to pin down. He'll tend to take an objective stance, and adopt neutral standpoints, whereby he'll distance himself from feelings. The ubiquity of air will be reflected in his widespread interests and his lively curiosity. And, as air cannot be easily contained, so the air person will need to feel that he is free.

Air stands symbolically for the rational mind and mental faculties; and an air person is often 'hung up in his head'. He'll go through life collecting items of knowledge that he'll organise into mental categories and systems of classification.

Air is also the element of communication, so the air person is likely to talk a lot. He'll feel a need to verbalise his thoughts and relay them to other people, and in today's world he'll probably be into the social media and network online with people across the globe. Air is the element that connects us with other human beings and all living creatures, as we all breathe the same air.

Finally we come down to earth. Feel the comforting solidity of your body which is your anchor, and move into a deeper understanding of the earth archetypes by contemplating images of rocks ... great tree trunks ... freshly dug soil ... metal objects ... beautiful crystals and gem stones. Imagine being in a cave deep underground, and sense the heaviness and thickness of the rock around you and above you ... Then what are earth's characteristics? (See Figure 25.)

Well, to begin with it's heavy and it's opaque. And, as it's a female element, which naturally moves inwards and downwards, earth will always sink – for example through water to form a layer of sediment at the bottom of a pool. It's the most enduring of the four elements and also the most structured, qualities which, when translated into psychological terms manifest as reliability and consistency.

The solidity of earth expresses in the earth person's need for security structures. She will choose a structured life style, prefer to plan ahead and strive to create firm material foundations. Her realism and practicality equip her well for mastering life in the material world. And her well-developed physical senses will mean she is more in her body than fire, air or water people are. However earth people can lack imagination, and be blind to the existence of other dimensions of reality beyond the material.

Your birth chart will indicate the proportions of the four elements in your make-up. To investigate their weighting you can start by counting the number of planets in fire, air, water and earth signs to find out which element is the most emphasised in this way. However you should then go on to include the elements represented by the houses occupied by planets, and the elements represented by the prominent planets in your chart, though some experience of astrology is necessary before this can be judged (see Appendix B).

Cosmic quadrupeds

The circle of the zodiac, as we have seen, contains not only four large triangles known as the *triplicities,* but also the *quadruplicities,* which are three large squares (Figure 13, p. 99) These are formed by joining the corners of the crosses that link the signs with the same mode of energy. Like four-footed beasts spanning the zodiac they then have one hoof in each *quadrant* (see Glossary).

As mentioned earlier, the modalities or modes represent three types of force – cardinal, fixed and mutable. These terms, rather than describing types of energy, describe the way the energies move. Below is a table showing the twelve astrological archetypes divided into the quadruplicities. The zodiac arithmetic behind them is as follows: three four-sided squares multiplied by four makes twelve sides in all, and four triangles multiplied by three also makes twelve sides.

Mode	Signs			
Cardinal	Aries	Cancer	Libra	Capricorn
Fixed	Taurus	Leo	Scorpio	Aquarius
Mutable	Gemini	Virgo	Sagittarius	Pisces

Whereas the energy of cardinal signs is dynamic and 'fast-forward', the energy of fixed signs is sluggish, as the forward movement is counteracted by a downward pull into depth. Mutable energy moves forward, but in doing so 'leaks' out on all sides; thus its momentum is dissipated. Therefore, in contrast to cardinal and fixed, mutable energy is unfocused and all over the place.

Just as the three modes were represented in each triplicity in the table of triplicities (p. 123), so in the above table the four elements are represented in each quadruplicity. For example if, looking back at Figure 14, (p. 100), we start at cardinal Aries in the fire triangle, and follow it counter-clockwise, we next meet fixed Leo and finally mutable Sagittarius. And similarly, if we start at earthy Taurus in the fixed square represented in Figure 13 (p. 99), and move counter-clockwise, we meet fiery Leo next followed by watery Scorpio and finally airy Aquarius. Following this pattern will give you a feeling for how these primary archetypal qualities relate to the geometry of the zodiac.

Thus the elements, which represent different states of matter, are incorporated with the modes, which represent different types of force. And together these zodiac triangles and squares produce the core meanings of the astrological archetypes, defining them both through similarity and contrast. Every cycle of events occurring in the macrocosm or microcosm must necessarily pass through the three phases represented by the modes, always in the same order – from cardinal to fixed to mutable and back to cardinal again.

The qualities of these modes characterising the successive phases of time cycles are on a cosmic scale. They govern the patterning of Nature's life cycles, and socio-political cycles in human collective life, as well as determining the rhythms of our individual lives. They appear as trinities in different theologies. In the Hindu, for example, Brahma as creator is an expression of cardinal energy, Vishnu as maintainer of fixed, and Shiva as destroyer of mutable energy. Below are three word fields that convey the contrasting sets of meanings of the three modes (see Figure 26).

As with the elements, your birth chart will also show the proportions of the three modes in your make-up, which you can discover in the way described on p. 128. If it turns out that you're a cardinal person, you'll be the one who initiates and sets balls rolling. You will have the original ideas at the start of an enterprise, and set off enthusiastically on your course, motivating others to join you. Beware of pushing yourself beyond your body's limits; your hyperactivity can lead to burn-out.

If you're a fixed person your role will be to provide an enterprise with structure and consistency. You've got the ability to stick at things, and persevere through challenging times because giving up is simply not an option for you. But beware of rigidity in your opinions and attitudes, which can lead to you getting seriously bogged down. And your tendency to resist change, even when it's long overdue, is your worst enemy.

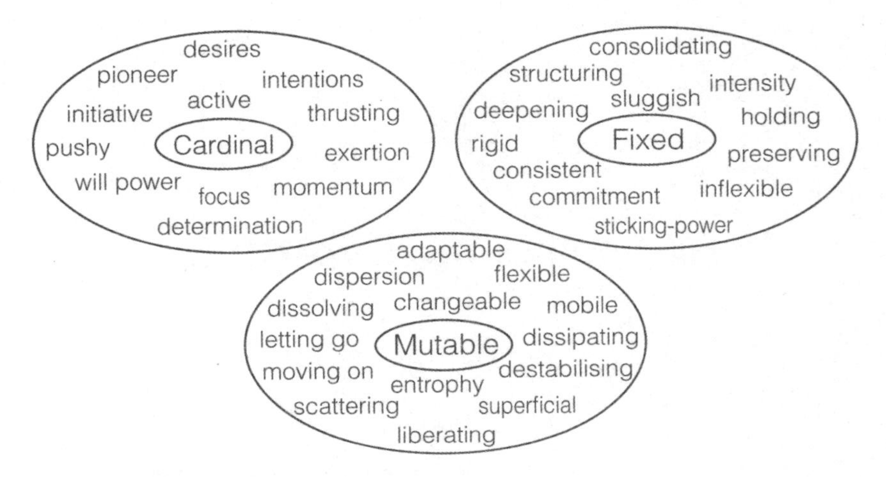

Figure 26. Word fields for the three modes.

If you're a mutable person your role in an enterprise will be to view things from a broader perspective. A widening of scope, however, can bring a decrease in forward momentum. Things become diffused and then confused, leading to a decline and an inevitable ending. But, as you're flexible and adaptable, you'll be able to take this process in your stride. Beware of moving on too quickly and of scattering yourself too thinly, or you'll brush the surface of life only and never savour its depths.

To sum up, all time sequences begin with a cardinal phase in which a new start is made. Then intentions are put out and initiating action is taken. This is followed by a fixed phase when what has been achieved so far is consolidated and deepened, which allows the status quo to be maintained for a while. However an unstable mutable phase necessarily follows, when what has been created is modified and weakened. Finally the momentum runs out, and what was built up during the cycle crumbles and falls.

However the mutable mode that brings endings is also liberating and necessary, as it clears the way for new beginnings in the cardinal phase to follow in the next cycle. Knowing about the sequence of these three phases can make us aware of the potential of any time period, and leads us to understand that energy is never created or destroyed; it simply changes its mode of expression.

CHAPTER 7

Archetypes in Horoscopes

The stars speak

'From the summit of the sky the stars speak. They know everything but compel no one. The wheel of time, formed with twelve spokes spins in the heavens maintaining order'.[1] The stars speak, but to discover their meaning we first have to learn their language. Down the long centuries astrology has preserved the language of the stars, but the modern mind, firmly encased in the scientific materialist paradigm, needs to be broadened in order to understand it.

Before starting out on an astrological interpretation, we need to be clear that the terms used in astrology apply to experiences on Level B (Fig.1, p. 22). This means the truth of a statement in astrology cannot be proved in the same way as the truth of a statement about an experience on Level A, because on Level B we are no longer inside consensus reality.

For example try explaining the meaning of a dream you had. Your feelings and imagination will become involved and introduce all kinds of stuff complexifying the facts – which doesn't prove your dream was not real. Meaning on non-material Level B is subjective and also holistic. It involves us as whole people with our feelings and imagination as well as our rational minds. To make this clear we can say that, for an astronomer, the meaning of the stars is found on Level A, whereas for an astrologer it is found on Level B.

Myths are also containers of Level B meaning. The archetypal patterns that play out in mythical narratives are dramatisations of the cosmic principles that inform creation. In them the twelve astrological archetypes make their stage appearances as characters, while themes from their fields of meaning combine to form the plots. There are just a handful of basic myths that play out in endless variations. And the general and timeless significance they have is explained when we understand that they stem from the archetypal strata within the collective psyche.

If our individual psyches are embedded in the collective psyche and thus are inseparable from it, it's no wonder we find versions of these myths playing out in our personal lives. Our birth chart will reveal which myths will be relevant in our case, with our planets representing the characters, and their placements and interrelationships the stories. As the cycles of our life unfold, these patterns are triggered by planetary transits, and then we find them being acted out.

If we compare a number of birth charts we'll find a different combination of astrological archetypes in each, represented by the planets in their signs, houses and geometric inter-relationships. The fact that each birth chart is unique means that endless variations on the basic themes of the main myths can be developed. And in Chapter 8 we'll look at William Blake's life, and see which myths characterising it are prefigured in his birth chart.

An archetypal approach to interpreting a natal chart recognises the archetypal patterning that is reflected in the biography of its subject, and sees this as an expression of the larger mythical patterns of the twelve astrological archetypes found in the cosmic psyche. These are the patterns tapped into, for example, by the authors of the Greek tragedies, and by Shakespeare in his immortal plays, and which serve to explain the timeless and universal appeal of these dramas.

Although we have located them on Level B, the meanings derived from a horoscope are not purely subjective and therefore arbitrary, stemming as they do from the archetypal layer in the cosmic mind. Nobody sat down in the remote past and decided that Mars would be the god of war and rule iron, weapons, butchers and surgeons. And neither is the association of Mars with warfare in astrology due to those Chaldean astrologers who sat up night after night recording how often Mars rising in the East coincided with war being declared. Rather Mars should be seen as integral to the cosmic archetype called Aries – one of the twelve archetypes that inform the natural world and the human psyche with meaning.

Therefore, although we may feed in some of our subjective associations when we interpret Mars in a chart, our work will be all the more meaningful the deeper we can tune in to the qualities of the cosmic archetype that the planet Mars symbolises.

Because of the essential oneness of the individual mind with the cosmic mind, an astrologer can contact the cosmic archetypes present in her own psyche in order to articulate archetypal meaning. And only if this is the case will her interpretation go beyond being merely subjective impressions or a list of ideas learned by rote. And then her findings will be objective to the extent that any other astrologer capable of accessing that deeper level of the collective psyche will agree with her.

What this means is that, to interpret a chart in any depth, an astrologer must merge her mind with the cosmic mind – a paradox as our minds are anyway fused with the greater mind being a part of it! Nevertheless it takes a focusing and an intensifying of our awareness to intentionally move down into the archetypal level of the collective psyche, like a bucket descending into a well, to scoop up some of its living water. And this focusing is not an effort or a narrowing but more like a letting go, an allowing of a merging with that deeper realm of consciousness to happen, which then triggers an expansion of understanding.

The mythologist Joseph Campbell wrote of the function of myth that it 'serves to reconcile the waking consciousness to the … universe as it is …,' and to 'open the heart and mind to the divine mystery that underlies all forms'.[2] And the same could be said of creating an astrological interpretation. It's said that astrologers can only recognise in another's chart what they've already recognised in themselves. I would add that an astrologer can only recognise archetypal meaning by accessing the cosmic dimension in the way described. And, each time she succeeds in aligning her mind with the cosmic mind, the act of interpreting a chart becomes co-creation with the universe.

Fields of buttercups and daisies

The idea of invisible fields that organise waves and particles of energy, like the electro-magnetic fields described in physics, was extended into biology by the biologist Rupert Sheldrake.[3] His 'morphic fields' serve to unify the diverse components of the systems of living creatures creating organisms, and as wholes these organisms are then greater than the sum of their parts.

Sheldrake's conception of morphic fields solves the riddle of how the billions of molecules in our body 'know where to go', by proposing that they are sent to their places and maintained there by embracing invisible fields. Thus our molecules are embedded in our cells, which are larger units also organised by fields. And these in turn are embedded in the fields that surround our organs, which in turn are governed by the largest field of all – that of our whole body. Taken together these fields great and small form a nested hierarchy, a concept we applied in Chapter 1 to minds (see p. 33). Therefore, parallel to Sheldrake's conception of invisible formative fields functioning on Level A, I am suggesting that similar formative fields organise the psyche on Level B (Figure 1, p. 22).

Imagine a single oak tree in the centre of a meadow surrounded by buttercups and daisies and other diverse flowers. The invisible roots of the old gnarled tree spread out in all directions under the soil in which the flowers are growing, holding them in its 'field'. In a similar way each astrological archetype can be imagined as magnetic nodal centre, attracting a cluster of themes that resonate with it, and are on the same wave length as the other themes in their group. It's as if the central archetype emits a frequency whose vibe attracts qualities, situations, events and experiences of the same genre.

Seen in this way the astrological archetypes appear similar to the 'attractors' of chaos theory, which chaos theorists describe as powers belonging to an order transcendent to that which they organise. And like attractors the archetypes transcend the subject matter they structure. Referring yet again to Figure 1, we can see an archetype as dwelling on Level C, and its field of meaning as a cluster of ideas, images and emotional patterns on Level B, which can then emerge – but must not – as events on Level A. In other words the contents of an archetypal field should be understood as latent but with a potential for manifestation.

The manner in which themes in the same field are related to the central archetype, and also connect with one another, can be compared to the fluid way that quantum wave patterns interpenetrate within the quantum field, or the way thoughts cluster round an idea in the mind, linking up with other associated thoughts. Remember Pauli's comment to Jung that the quantum field appeared to him very much like a giant mind?

Jung must have been thinking in this way when he developed his conception of archetypal complexes composed of beliefs, images and stories that are associated with the same archetype. His archetype of the Mother, for example, is described like a node around which a complex of images and associations cluster to do with things like home, nurturing and childhood. In my view, however, the Jungian archetypes are subsumed by the astrological archetypes. Cancer, for example, includes all the ideas ascribed by Jung to the Mother and more.

It is important to remember that, within an emergent universe (a conception now being advanced by progressive cosmologists), archetypal fields will necessarily be in flux, and their range of meaning should be seen as open-ended. For this reason some of the themes assigned to Cancer in astrology today may not have been part of the Cancer field in the past – the idea of 'emotional co-dependency', for example, which entered it with twentieth century psychological astrology.

Scientists are now discovering that not even the laws of nature are as permanent as we once thought, which makes sense when the universal mind is understood as a process rather than a thing. Therefore astrologers should be aware that no meanings ascribed to the astrological archetypes that go beyond numbers and geometric figures will ever be ultimately true. Meanings in astrology can and do mutate. And consequently no astrological approach or system of interpretation, and many have been devised over the centuries, will ever be the only true one.

Teeth, goats, ivy and the Arctic

In theosophical teaching the zodiac principles are described as twelve rays of the infinite that manifest throughout all creation, which is a beautiful and apt metaphor. We can imagine them as coloured strands weaving themselves into a tapestry and producing all the varied hues of creation.

A thing or a phenomenon can be understood as having a particular quality or meaning because it participates in a certain archetype. This is the received wisdom lying behind the medieval concept of the chain of being as a system of correspondences. For example the items in the Aries field, as illustrated in Figure 27, cover the animal, vegetable and mineral kingdoms on Level A, as well as human personality traits and propensities on Level B. The

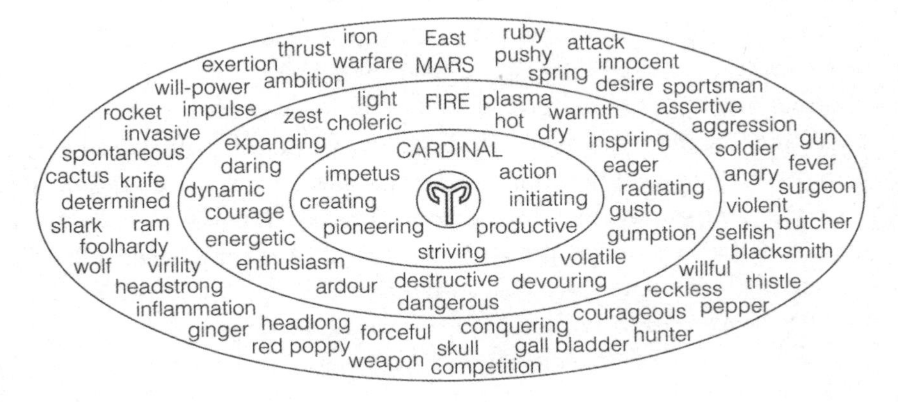

Figure 27. The Aries archetypal field.

thistle as a plant, the ram as an animal and the warrior as a human type all participate in the Aries archetype. And it's no coincidence that thistles, rams and warriors are tough and horny as they have an affinity with weapons, iron and the planet Mars!

Similarly a sunflower in the realm of flora and a lion in the realm of fauna both participate in the Leo archetype of which the sun is the planetary representative. So it's understandable that both sunflowers and lions remind us of the sun with their ruffles of petals and tousled manes like the sun's rays. The golden colour they share is also explained by their affinity with the metal gold. And these correspondences, which sound cute and naïve like something from a child's picture book, can be explained by archetypal participation.

Archetypal fields lie behind the complex system of sympathies between planets, metals and plants that from ancient times has been central to astrology as well as to medicine and alchemy. And learning astrology involves in the first place becoming familiar with the contents of these twelve categories.[4]

To recognise the archetypes as they manifest in our daily lives, we need to think 'vertically', rather than horizontally, as we are used to doing, in terms of cause and effect. For example, if I asked you what gold, silver, copper, iron and lead have in common, you would reply they are all metals. And so, I would continue, what do teeth, the mountain goat, a coalminer and the Arctic have in common? If you knew some astrology and you could think vertically, you'd reply they are all expressions of the Capricorn archetype.

This second type of list links items through analogy rather than through causal thinking. Things belong together because they participate in the same archetypal field.[5] When we're awake we usually think horizontally, but we think vertically when we're asleep. Thus our teeth, a coal miner, ivy and the Arctic could all appear woven together in one of our weird dreams.

I've discovered that the most effective way of communicating an archetype's qualities is graphically, as demonstrated in my diagram of the Aries archetypal field (Figure 27) as well as in the other eleven field diagrams you will find in Appendix E. It's a right-brained rather than a left-brained approach to them.

A diagram is more appropriate for exploring the meanings of a set of symbols than a paragraph of writing that comes mainly from the left brain. It is also in line with the method of amplification used by Jung when he was working with archetypes. Linked visually, as in Figure 27, the words in the same field gain significance through their spatial relationships, and define each other through their complementary meanings.

The keywords I've included here, and in the other diagrams in the series, are a personal selection of possible words that could be used to signify the archetype's meaning. However I believe the most important ones have been included. They are arranged in order of significance from the centre outwards. Thus the words closer to the centre of the diagram manifest more fundamental qualities and themes related to the archetype than the words closer to the periphery.

The two concentric circles surrounding the central glyph contain words for the qualities of the element and mode of the archetype. The same words will also be found in the central areas of the field diagrams of the other archetypes that share the same element or the same mode.

At the heart of each diagram stands an astrological glyph representing the ineffable core meaning – 'the true name of the god' as the ancients would say. I see the twelve zodiac glyphs (see Appendix A) that we have inherited from remote antiquity, and which in astrology are used to symbolise the signs of the zodiac, as visual representations of the astrological archetypes *per se*. Try meditating on them, because their shapes that mirror the specific energy flow within each archetypal principle are keys to the archetypes' core meanings.

Twelve divisions of the sky

So what exactly are the signs of the zodiac? In the first place they are not the same thing as the astrological archetypes. The archetypes are the ineffable, numinous essences represented in a chart not only by the signs, but also by the houses and planets. Thus we find the Aries archetype manifesting its typical qualities in the sign Aries, in the first house and in the planet Mars. Figure 28 shows how signs, houses and planets would match up in a chart where all these three factors aligned, although the chances against this happening must be trillions to one!

The diagram in Figure 28 is composed of two wheels each divided into twelve sections. The segments of the outer wheel are distinguished by the zodiac glyphs representing the signs, and

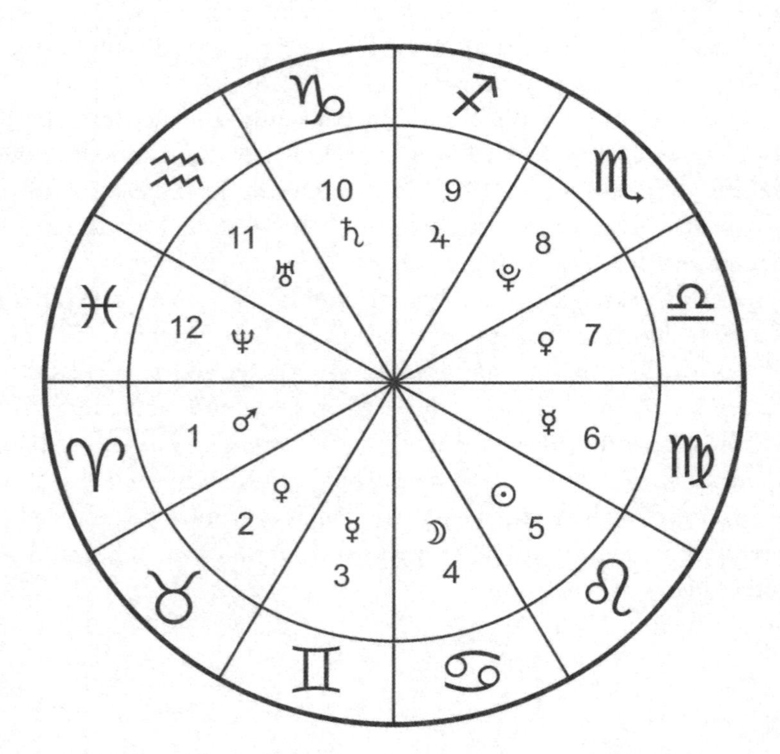

Figure 28. The twelve archetypes matching signs, houses and planets.

those of the inner wheel that are numbered 1–12 represent the houses. In archetypal astrology the two wheels mirror each other on the quantitative and qualitative levels of their geometry.

The planets in Figure 28 have been placed in the signs and houses corresponding to the archetypes they belong to, or 'rule', whereby you'll notice that Mercury and Venus occur twice. They each rule two signs and houses. We can infer from the diagram, for example, that the planet Neptune, the sign Pisces and House 12 all express the archetype Pisces.

Only once in twenty-four hours is the state of perfect alignment between houses and signs, that is shown in Figure 28, briefly reached. That's when 0 degrees Aries on the outer wheel aligns with the *cusp* (see Glossary) of House 1. At all other times of the day and night the two wheels are out of alignment.

The signs of the *tropical zodiac,* (see Glossary) which I am using here, are twelve thirty-degree divisions of the ecliptic. So they are not the same thing as the twelve constellations of stars that form a backdrop to the ecliptic, and which, confusingly, have the same names as the signs. It's very likely that these twelve constellations were named after the astrological archetypes rather than vice versa as is supposed.

The psychological characteristics that astrology attributes to the twelve zodiac character types can be explained by their different combinations of gender, element, mode and ruling planets (see Appendix B). Each sign of the zodiac is a possible permutation of a polarity, a triplicity and a quadruplicity, as we have seen in Chapters 5 and 6, and the meanings these combinations engender therefore derive from the geometry of the zodiac.

For example to understand the psychology of the Gemini type, we would combine the concepts of active and extravert (gender: male, positive), communicative, mentally and socially oriented (triplicity: air), flexible and adaptable but changeable (mode: mutable), and add the mythological associations of Mercury, the ruler of Gemini, (planet: magician, alchemist and winged messenger).

And now, if you possess a copy of your birth chart (and you

can get one free online today), you can contemplate the qualities of your sun sign, moon sign and rising sign along these lines, which represent the three main archetypes that are active in your psychological make-up.

Twelve divisions of the earth

So what are the houses exactly? Take a look again at Figure 28 where the sky is represented by the outer circle of zodiac signs and the earth by the inner circle of the houses. The houses are divisions of the local landscape around the place for which the chart has been drawn up, and therefore 'ground' the patterns of the sky, formed by the planets, by applying their meanings to a particular time and a particular location.

If we could animate this diagram, we'd see the wheel of the signs turning slowly clockwise round the wheel of the houses, taking twenty-four hours to complete a full circuit, mirroring the earth revolving on its axis. While it turns we'd observe the sun, moon and planets which it carries rising in the East at the Ascendant, culminating in the south and descending in the west at the Descendant.

The convention of the houses representing twelve divisions of local space possibly originated in the series of sighting points along the horizon that Neolithic astronomers would establish for time-keeping purposes. By the means of these landmarks they could tell the times of the day or night by measuring the passages of the sun or stars in relation to them. They would also determine the season of the year by recording the slow drift of the sunrise point along the eastern horizon.

The house cusps are therefore like sight lines radiating outwards from the centre of the circle to mark twelve compass directions, which, if they are extended upwards at the horizon, also divide the sky into twelve sections (see Figure 29).

Over the centuries a number of mathematically and geophysically talented astrologers, with a bottom-up, rather than

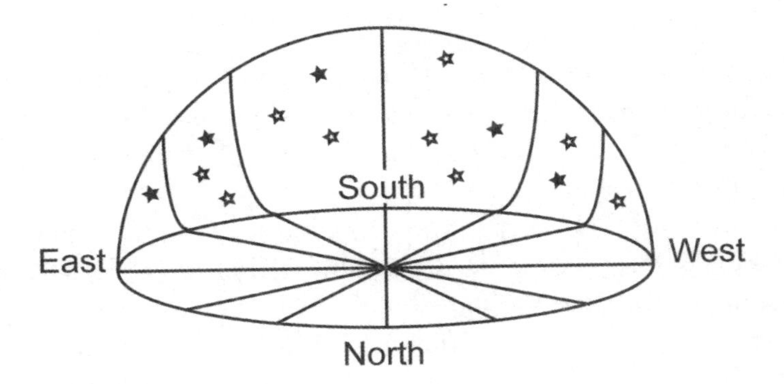

Figure 29. The origin of the houses.

a top-down, approach to the problem, have experimented with different ways of dividing the inner wheel of the houses in a chart. Thus different house systems have arisen, and today the discrepancies between the systems used by different schools of astrology can confuse beginners.[6]

Archetypal astrology, however, favours the Platonic top-down approach to house division by which the inner wheel of a chart is divided into twelve equal sections of thirty degrees. It results in a chart where both houses and signs derive from the same geometry, are based on the same numerological principles, and both wheels reflect the symmetry of the zodiac matrix.

Have a look at the alignment of the two wheels in your natal chart, which will reflect the relationship between earth and sky at the time and place where you were born. Hopefully the birth time used to calculate your chart is close to the moment of your first breath, because even a small error will make a difference. As every two hours, on an average, a new zodiac sign rises on the eastern horizon due to the earth rotating on its axis, a wrong birth time can produce a wrong Ascendant, and the Ascendant, as we have seen represents one of your main archetypes.

The many possible alignments of the outer and inner wheels in a chart open up a whole range of potential archetypal combinations, and produce a variety of personality types. For example there are twelve types of Virgo: the Virgo–Aries combination when Virgo is the rising sign and therefore on the first house, and the Virgo–

Taurus combination when it's on the second house, and so on. This implies that a lot of meaning can be gleaned from the way the two wheels in a chart are aligned even before the planets are taken into consideration.

In an archetypal interpretation each house contributes qualities and subject matter from its corresponding archetypal field to the meaning of the sign it's aligned with. Thus a chart with the sign Capricorn on the cusp of the first house (the house naturally ruled by Aries, see Figure 28), will produce a combination of qualities from the Aries and Capricorn archetypal fields which will characterise the personality of its owner.

Such a person would, for example, be strong willed and dynamic (Aries and Capricorn are both cardinal archetypes), as well as fiery (Aries) and earthy (Capricorn) at the same time – which could be a problem! Fire qualities and earth qualities are very different. However this combination can turn out to be productive when the earth qualities provide a solid base for the fire, while setting it necessary limits, and the fire qualities inspire and energise the earth.

As the same order of the twelve archetypes that structures the wheel of the signs is repeated in the wheel of the houses, its sections should be interpreted similarly.[7] However, within the subject matter of the houses there is an additional emphasis on conditions and events in twelve specific areas of life (see Appendix C) which are particularly relevant to the interpretation of planetary cycles.[8]

Within us are sun, moon and stars

The planets are often referred to as archetypes, but I believe a distinction should be made between the twelve astrological archetypes as numinous essences, and the planets as their visible representatives. Planets are objects in the physical world, and as such belong on Level A of Figure 1, whereas the archetypes, as we have seen, dwell on non-material Level C. The main planets used in modern astrology are the ten whose glyphs are illustrated in Appendix A.

In the ancient world the planets were seen as gods and worshipped as such, yet again they are not the same thing. Divinities are personifications of aspects of archetypes that have become autonomous identities within the cosmic mind. They are maintained by the intensity of collective belief in them. The character traits ascribed to the Roman gods, after whom our planets are named (excluding those more recently discovered) reveal the astrological archetypes that engendered them. For example Saturn is a clear manifestation of the Capricorn archetype, and Jupiter reflects the qualities and themes of Sagittarius. In my opinion there is nothing eternal about gods and goddesses; they all have their day; they rise to power and then fade from view – or morph into new deities.

The myths that have come down to us from antiquity depicting the exploits of the gods are widely used in psychological astrology to define the astrological meanings of the planets. Though this does not imply that the 'characters' of the planets are arbitrary projections from the world of myths. Instead it points to their archetypal source.

The extensive study of the relationships between planetary positions and world events over many centuries carried out by Richard Tarnas, has demonstrated the extent to which the archetypal themes of the planets are reflected in the political, social and cultural events of different historical periods.[9] As an archetypal astrologer Tarnas does not see the planets as causing these events. Rather he sees their cycles as reflecting processes in the cosmic mind which become mirrored in events on earth. They are like ideas that are being worked out over longer periods of time and in different sets of circumstances.

Each planet is a quintessence of the meanings of the archetypal field to which it belongs, and therefore they should be seen as carriers of the qualities and themes of the archetypes. And, as the cosmic mind is an ongoing process and its ideas are in flux, there is flexibility in the relationships between the two. Several planets, for example, can represent the qualities of the same archetype. Thus Saturn (as its old ruler) and Uranus (as its new ruler) both represent facets of the archetype Aquarius. Conversely more than one archetype can be represented by the same planet, as in the case of Mercury, which acts as a vehicle for the qualities of both Gemini and Virgo.

Within this elastic system newly discovered planets such as Chiron and those recently found in the Kuiper Belt can be included in an astrological interpretation without astrologers having to find new archetypes to accommodate them. By the way, astrology has no problem with Pluto being demoted to a dwarf planet by the astronomers. However they choose to define Pluto, in astrology for the present at least he seems to be functioning as the main vehicle for the Scorpio archetype. The archetypes are eternal, but their planetary vehicles are like hired bicycles – used for a short time and then left standing at the kerb.

To return to our imaginary animation of Figure 28, while the outer wheel is turning clockwise round the inner in the course of twenty-four hours, the moon moves counter-clockwise round the outer wheel taking a month to complete a circuit. Meanwhile the sun also moves in the same direction through the signs at the rate of one sign a month, while the other eight planets follow, taking varying times to complete their circuits depending on their speeds and distances from the sun. Mercury is the fastest, and Pluto the slowest.

As they move round the zodiac, the planets carry the subject matter of the archetypes they represent into each sign and house in turn. These transits temporarily expand the meaning of these areas of the chart, adding the visiting planets qualities and associations to the archetypal mixture already present. For example the sun that rules Leo carries the themes and qualities of the Leo archetype through all twelve signs in the course of a year, whereby each sign receives Leo's gifts differently according to its nature. In this way the planets serve as a third source of archetypal meaning in a chart interpretation.

'Understand that you are another little world and have within you the sun, the moon and also the stars', wrote the third century theologian Origen.[10] I can relate to that idea. When the outer sun rises in the morning my inner sun also peeps over the horizon and I awake. When the outer moon is full in the sky, my inner moon is also full and I brim over with feelings. It's not a case of the planets influencing me in any cause-and-effect way. Rather it's my daily proof of the truth of the Big Secret – that the microcosm and the macrocosm mirror one another!

Walkabouts through astrological landscapes

The aspect lines that link up the planets in a chart are derived from the interior geometry of the zodiac. For example we've found six polarities (oppositions), three large crosses (squares), and four large triangles (trines) that link the signs in different geometric figures. If we include the twelve small triangles around the periphery of the circle *(sextiles)* and the 150° aspects *(quincunxes)*, that covers the main set of aspects used by astrologers today when they interpret a chart (see Glossary and Appendix D).

The word 'aspect' derives from the Latin word meaning 'to look at', and the zodiac signs look at each other in qualitatively different ways depending on the angles between them. For example they look antagonistically at signs lying 90° away, whereas they smile at the signs that make a 120° angle with them. In the former case there's a conflicting clash of energies, and in the latter easy, harmonious relating. It all depends on the maths! Finally it's the internal geometry of the zodiac that decides which signs harmonise and complement each other, and which ones clash and repulse one another.

Plato's five polyhedra (see p. 62) form the geometric basis of astrological aspects. For example squares are sides of the cube, and trines are sides of the tetrahedron. As we have seen, Plato assigned an element to each of his polyhedra, whose qualities we can now transfer to the respective astrological aspects. The earthy qualities, for example, describe the nature of squares, and the fiery qualities the nature of trines.

Changes in the physical world were described by Plato as transformations of energy from one element into another, whereby the geometric solids were broken up to then reform. Translated into psychological terms this reflects the process of 'inner alchemy' by which, for example, we move away from behaviour associated with the intractable squares in our chart towards the aspiring and growth-oriented perspectives of our trines.

So we can see the aspect lines as tracing the flow of energy in a horoscope. However, in the case of squares, this flow is held up by the 90° angles which act as bottlenecks. You'll never find a plumber laying one water pipe at right-angles to another! Square aspects hold things up causing tension and frustration, though they can also trigger dynamic action in response. As a geometric shape the square reflects the limits set to our boundless desires by element earth with its non-negotiable facts and rigid structures. When combined in an aspect pattern with a trine, however, a square can ground the trine's aspiring arrows of desire, and make the growth it triggers sustainable.

In contrast to the square the triangle is composed of 120° angles through which energy passes in an easy flow, and which serve to boost its momentum. The tetrahedron is surprisingly a more stable form than the cube, which is why milkmaids' stools traditionally had three legs – to prevent the cows from kicking them over! Have you noticed how a candle flame adopts a triangular shape when it burns? And, just as the triangle corresponds to the element fire, trines are associated with ascension and aspiration in astrology.

The hexagram or six-pointed star is the basis of the sextile aspect. There are two hexagrams in the zodiac – one formed by the male signs and the other by the female. Sextiles are pathways that harmoniously link the six signs within each of these two groups. This shape is a basic element in the geometry of the sri yantra mandala, which depicts upward pointing yang triangles in perfect equilibrium with downward pointing yin ones.[11] When combined in this harmoniously symmetrical image they symbolise a *unio mystica* or mystic marriage between the male and female principles. In other words your sextiles are pathways that can lead you towards divine union!

The patterning created by the aspect lines in your chart is your personal sacred geometry. It will reveal which archetypes are linked in your case, and so have the power to mutually activate one another. It will also show which archetypes are merged, as in the case of *conjunctions* (see Glossary), to produce complex compound meanings. Your aspect patterns will point to areas of harmony as well as areas of discord in your life, revealing the sources of your self-defeating patterns and indicating your growth points.

An archetypal approach to aspects sees them in the context of the geometric patterns inherent in the relationships within the circle between the twelve zodiac principles. When interpreting their meanings in a chart, remember that the nature of each – whether a harmonious trine or a dissonant square for example – is secondary to its function as a pipeline. Aspect connections allow information from archetypal fields to flow into one another and so produce new combinations of meaning.

Looking at the birth chart as a map as it is sometimes called, the aspect lines become pathways inviting us to go on a walkabout through an astrological landscape. Some of the paths will be dead-ends, with the only option being to move backwards and forwards along them. Some will lead to conjunctions, which can act like bridges for crossing from one planet to another adjacent to it. And from there we can often set off on a new route round another geometric shape.

A walkabout of this kind can help us become aware of the disparate areas of our personalities and of our lives, and help us integrate them. Then we are using astrology for its highest purpose – namely for spiritual development and the growth of consciousness.

And, if you set off on a walkabout, you may discover something new in your chart you've never seen before. For example you could, like me, discover that you have one of those very potent Pythagorean triangles consisting of a square, a trine and a quincunx. In the ancient world, as we've seen, this 3:4:5

triangle was special because it had mystical associations. These had to do with this type of triangle being used to square the circle, metaphorically uniting heaven and earth and translating the manifested back into the unmanifested. The numbers of the Pythagorean right-angled triangle also generate spirals or vortices, which Neolithic shamans could have used to create routes for ascension in their astral bodies through the dimensions.

Finally an exploration of the aspect pathways in your chart will make you aware of the geometric core of astrology that mirrors the geometric underpinning of the cosmos.

CHAPTER 8

Archetypal Astrology in Action: William Blake's Chart

Till we have built Jerusalem

In this chapter we demonstrate the truth of the 'big secret' of the ancients – that the microcosm and macrocosm mirror each other, and the patterns astrologers identify in the sky at a particular moment correspond to the soul patterns of an individual born at that time.

Figure 30 is a map of the sky above London on the November evening when the artist and poet William Blake came into the world – in other words it's his birth chart.[1] It shows a particular alignment of the wheels of sky and earth, and the positions of sun, moon and planets in relation to the horizon and to one another. Taken together these factors reveal Blake's inborn personality patterns, and suggest the kinds of experience he'll consequently 'attract' in his life.

At 7:45 pm on November 28, 1757, the last degree of Cancer was rising on the eastern horizon, giving Blake a Cancer Ascendant close to the Leo cusp. As a consequence his Descendant is twenty-nine degrees Capricorn, with his MC and IC at four degrees Aries and Libra respectively. Blake's main aspects are represented by the geometry in the diagram's central area, with planets in opposition and square aspect joined by thick continuous lines, and those in trine and sextile by thin continuous ones. The dotted lines indicate quincunxes. For simplicity's sake aspects between planets and cardinal points have not been included.

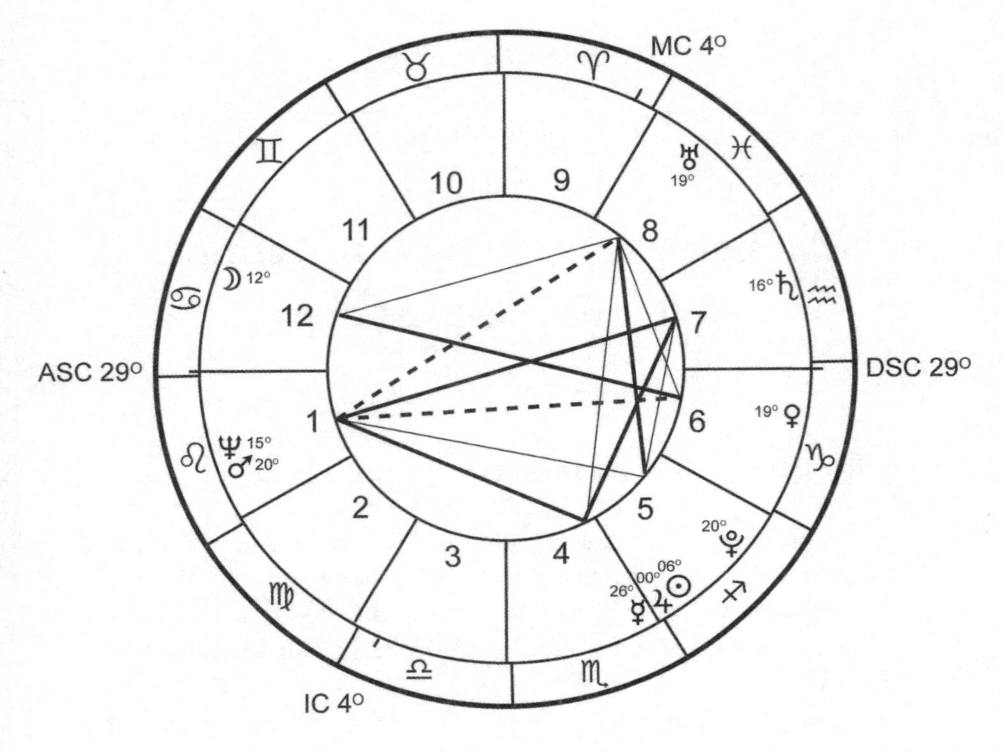

Figure 30. The birth chart of William Blake.

Of course this same chart would apply to any other individual born at that hour on that day in that part of London. So, you might ask, as other babies must have been delivered then and there, how come they didn't all become famous poets like Blake? Because the astrological archetypes are multivalent and multi-dimensional, and therefore their ranges of meaning can manifest on different levels and in different ways – though archetypal coherence, as Tarnas calls it, always ensures that an archetype's core meaning will be consistently present through all its variety of forms.[2]

Also, as life is co-creativity with the archetypal powers that be, Blake himself had some say in how his archetypal potential manifested. The astrological archetypes are vast cosmic forces flowing through all that is, and they use us as conduits. Nevertheless our personal will power and our intentions, both conscious and unconscious, play a role in determining how the bare bones of our birth charts will be fleshed out into a human character and life story.

We are all from the same mould, being composed of the same twelve ingredients (the astrological archetypes), yet each of us is unique. Did you know that no one will ever be born again with exactly the same birth chart as yours? It's a statistical fact! The range of possible combinations of the twelve archetypes, as represented by the signs, houses, planets and other astrologically significant factors, is too great to allow this to happen!

We're like cakes that have been baked using twelve ingredients – the archetype we call Virgo could be the flour, Cancer the butter, Taurus the sugar, Capricorn the salt and Sagittarius the baking powder. And, as every baker knows, changing the proportions of the ingredients of a cake recipe even very slightly alters its consistency and flavour. Thus it's the differences in the proportions of the archetypes within us that account for the multitude of variations on the theme of the cake of human character.

When I was choosing a famous person's chart to serve as an example for this chapter, synchronicity struck again. William Blake first came to mind because we have a reliable birth time for him, his chart having been published during his lifetime.[3] Back in the 1960s I'd also written an MA thesis on Blake's reputation which I could revisit for material. But it was only when I started to re-read Blake that I realised what a suitable choice he was, because Blake was heavily into archetypes. His paintings and poetry were all about the elemental creative powers of the cosmos that he experienced through visionary insight.

Blake also fits in with the subject matter of this book because he was one of the earliest thinkers to reject the worldview of scientific materialism. He spurned rationalism and the values of the Enlightenment, valuing instead intuition and the powers of the imagination. He also claimed that his inspiration came from 'a body of secret knowledge from before the flood', and in his art emulated the 'lost originals from the ancient pre-Greek cultures' that he saw in vision.[4] Finally I discovered that the same Uranus-Pluto square presently in the sky (in 2013) was the major planetary configuration when Blake came into the world.

Opposition is true friendship

William Blake was born into a tradesman's family who lived in Soho where his father had a hosiery shop. He had one sister and three brothers. However the only family member he was close to was Robert, his youngest brother who died in his teens. Blake nursed him through his last illness, and afterwards claimed that Robert would assist him from the other side. 'I hear his advice and even now write from his dictate,' he revealed.[5] Blake, as this comment reveals, was a psychic who could see the spirits of the dead as well as other non-corporeal entities.

His artistic talents were recognised while he was still a child, and he was sent to drawing school before being apprenticed to the engraver James Basire. Basire would send him out to copy the effigies on the tombs in Westminster Abbey and other London churches, which is where Blake acquired his taste for the Gothic style in art. He'd been taught to read and write, and continued educating himself, learning Latin, Greek and Hebrew and reading widely in the classics and in works of philosophy.

His apprenticeship over, and having ambitions to become a painter, he enrolled as a student at the Royal Academy where he was taught by the likes of Sir Joshua Reynolds. Blake, a born rebel and inveterately intolerant of authority figures, objected to Reynolds' style of painting and fiercely criticised the values he represented. His stance of protest, and his refusal to be humble in order to curry favour, would be detrimental to his career as an artist.

He married a simple, illiterate woman – his dear Catherine

– and taught her to read and write and work side by side with him as his assistant. It was a happy marriage providing Blake with the support and security he needed till the day he died – although there are some reports of a stormy patch when Blake started advocating free love.[6] We'll decide whether his biographer Peter Ackroyd was right, and it all remained 'a matter of the mind rather than of the flesh' when we look at Blake's chart.[7]

In 1783 he published his first book of poems. Blake was an ardent republican at that time, hobnobbing with the group of London dissidents and radicals that included Tom Paine of *Rights of Man* fame, and the feminist author Mary Wollstonecraft. We hear of him being invited to read his poems at their London meetings. However the association did not last – Blake was not a joiner of groups but a maverick who needed to go his own way and do his own thing

His paintings were still being accepted for exhibition at the Royal Academy at this stage, and he was ambitious to become a great artist. But his style was not in line with current taste. It was too wooden and Gothic, and he liked painting wild, sublime subjects, such as ancient druids on rocky shores, which were seen as unsuitable for drawing-room walls. Also Blake was dependent on the services of a printer and bookseller to sell the work he produced, and there lay the rub, because he was too absolute in his demands and too scathing in his judgments to get on with most people. So, apart from a handful of patient friends and patrons who stood by him, his business relationships soon foundered.

In order to be independent in this respect he invented his own original engraving technique that enabled him to print his books at home in a cottage industry set-up. The first work produced by this method was the *Songs of Innocence and of Experience*, its pages being printed off one by one and then hand coloured. It was a work of high originality in its content as well as its production process, with the words and illustrations designed on the pages to create composite meanings. The many poetic jewels it contains include *The Lamb, The Chimney Sweeper* and every schoolchild's favourite – *The Tiger.*

From that time onwards Blake and his wife toiled unceasingly at engraving, printing and colouring the numerous books he went on to write and illustrate, which included twenty long epic poems – his prophetic books. Today they fetch enormous sums at Sotheby's, but while Blake was alive they never reached a wider public. A handful of copies were purchased by artists and connoisseurs, while Blake continued to earn his meagre crust as a hack engraver, producing book illustrations on commission designed by artists far less talented than himself.

In 1800 the Blakes left London for the first and only time to spend three years in a cottage on the Sussex coast. It had been offered to them by the poet William Hayley in exchange for engraving work. At first Blake made a great effort not to upset Hayley – he badly needed his patronage! However the problem was he didn't think much of Hayley's poetry, and found his own far superior, but holding his tongue went against his grain.

After months of suffering Hayley's benevolent despotism and growing increasingly frustrated, Blake exploded. His rage, however, was not directed at his patron but at a stranger, a soldier he'd discovered loitering in his garden. According to reports Blake abused the man verbally and removed him by force, and as a consequence was charged with assault and sedition. Apparently he'd damned the king and all his subjects, which was unwise with the country at war with republican France, and an invasion alert in force along the Sussex coast!

While waiting for his case to be heard at Chichester Assizes Blake suffered acute anxiety, fearing his past connections with the London republicans would be discovered. If it came out that he'd supported the Jacobins, and walked the streets wearing the *bonnet rouge,* he could be branded a traitor and thrown into gaol. Luckily it didn't come to that. The lawyer Hayley hired to defend him persuaded the judge that Blake was a harmless eccentric rather than a national threat, and so he was acquitted. However the irony is that many things Blake wrote *were* seditious. He hated kings and wanted them all deposed. However the recondite symbolic language of his writings, which makes them obscure to most people, was camouflage enough.

After this misadventure the Blakes returned to London and a period of great hardship began. Blake, urgently in need of commissions for work, was passed over again and again in favour of rivals. He reacted with proud defiance. For example when the artist Thomas Stothard won the commission Blake had coveted to paint a large prestigious picture of the *Canterbury Pilgrims,* Blake painted a rival picture on the same subject (considered today far superior to Stothard's) and exhibited it together with other work in a private exhibition set up in his brother's hosiery shop. Attendance was poor, and none of the works exhibited were sold.

Following this defeat Blake retired increasingly into his inner world and the private space of his home and workshop. The engraved plates of his books mounted up on the shelves awaiting orders that never came. He would comfort himself, however, that he was laying up treasures in heaven, because he believed that all the creations of every poet and artist are preserved in the halls of eternity.

It was only through the charity of friends and the thrift of his wife who, we're told, always managed to find something for the table, that the Blakes survived these years. By 1822, however, their straits were so dire that an appeal was launched at the Royal Academy 'recommending to the charitable consideration of the council Mr William Blake, an able designer and engraver, labouring under great distress'.[8]

It was around this time that Blake met the astrologer John Varley, who was interested in Blake's mediumistic skills. Varley organised 'sittings' for him during which Blake would summon up the spirits of the famous dead. Richard Coeur de Lion and the 'man who built the pyramids' are reported to have been among those who appeared to pose for their portraits. There are also reports of Blake arguing with Varley about astrology, but unfortunately no records of his views on the subject have survived.[9]

Then towards the end of his life a miracle happened. Blake had always been too far ahead of his time to be understood, but now the times were beginning to catch up with him. A group of young artists in their twenties discovered him and made him into their guru.[10] They not only recognised his genius and appreciated his work; they also rescued him from his isolation. He was invited to visit them and enjoyed their company. They particularly admired his selfless devotion to art in the midst of poverty and neglect, and emulated the simplicity and spirituality of his style of painting.

Blake died in July 1827 of a gallstone infection, and, according to those present, his death was amazing. He spent his last hours sitting up in bed singing, and, as death began to creep up on him, spoke of the spiritual wonders he saw, 'finally resigning himself to his eternal rest like an infant to its sleep'.[11] Catherine claimed afterwards that his spirit remained at her side, and that she consulted with Mr Blake on every decision.

Two contrary states of the human soul

In order to interpret a chart archetypally we must allow more space to the right side of the brain with its capacity for holistic intuition. Then our inner eye can glimpse the ripples of association expanding from the symbols and flowing together to create a composite meaning. The archetypal approach to interpreting is top-down rather than bottom-up – more in line with the approach to life of a Pythagorean mystic than a reductionist scientist. Thus we begin by viewing the distribution of the astrological archetypes within the chart as a whole in order to 'weight' them.

We find most of the planets in Blake's chart in the female night hemisphere, with only the moon, Uranus and Saturn above the horizon in the day half representing the outer world and public life. This weighting emphasises Blake's inner world and his private, domestic sphere. There's also an emphasis on the western side of the chart with an accumulation of planets in Houses 4–7. In astrology, as we have seen, the eastern side with the Ascendant symbolises Self, and the western side and Descendant the Other. This weighting tells us therefore that Blake was particularly sensitive to how other people reacted to him.

A birth chart is a diagram of a 'human form divine' to use Blake's famous oxymoron. It shows the twelve divine archetypal powers combining to create the particular mix of a human individual. In every chart four or five archetypes will be represented more strongly than the others, and these will usually be symbolised by the Ascendant and the Sun and Moon, plus the signs corresponding to their houses – in that order. In Blake's case these are the Cancer, Sagittarius, Cancer, Leo and Pisces archetypes respectively. Thus Cancer is strongly represented, being the sign of both Moon and Ascendant, but as Jupiter, ruler of Sagittarius, is in his own sign and in conjunction with the sun, I would weight the Sagittarius archetype as equally strong in Blake's case.

The next step is to investigate the proportions of the four elements in the chart as a whole, remembering that the elements together with the modes and polarities derive from the qualitative level of zodiac geometry in a top-down flow of meaning. In Blake's chart we find all three fire archetypes represented (his first-house Mars bringing in the Aries fire), which makes him a pronounced fire type. However we should note the presence of two outer planets in fire signs, namely Neptune and Pluto, as these could indicate potential problems in Blake's expression of his fire.

The three water signs are also occupied by planets, forming a grand water trine when wider orbs are taken that mirrors Blake's powerful emotions and strong imaginative powers. However, once again, we find an outer planet in water – Uranus in Pisces – signifying potential emotional turbulence and instability. Air, the element of reason, needed to balance all this water, is present only in House 7, and represented by Saturn in Aquarius – another outer planet here indicating potential problems in the area of Blake's relationships and communication. And earth, which Blake badly needs for his grounding, is only brought in by Venus in

Capricorn in House 6, suggesting he'll rely on his wife Catherine to do his grounding for him!

Looking next at how the three modes are weighted in Blake's chart, we find four planets in mutable signs including the sun in Sagittarius. This is surprising as in his biography Blake comes over as very fixed in his lifestyle and habits. But the mutable planets in his chart are all in fixed houses, and, as he also has four planets in fixed signs, the preponderance of the fixed mode that Blake expresses can be explained. Cardinal energy is provided by his rising sign, his Moon in Cancer and by Venus in Capricorn in opposition to it.

Next we weight the contribution of the individual planets, seeing each as a carrier of the qualities and meanings of the archetype it represents. The more prominent the planet the greater its contribution will be to the overall picture. For example planets in the first house, such as Blake's Neptune and Mars, are prominent, and also planets in conjunction with the sun such as Blake's Jupiter and Mercury. In general we estimate the weight of the contribution that each planet makes to the overall meaning of the chart by judging its strength in the context of the gender, element and mode of its sign and house.

The table shown opposite summarises my archetypal weighting of Blake's chart. Sagittarius, Cancer, Leo and Pisces emerge as his main archetypes, followed by Scorpio, Capricorn and Aquarius, whose signs are occupied by planets, and finally by Libra, Virgo and Aries whose corresponding houses are occupied.

Neptune's strong position in House 1, which brings in additional Pisces energy, is noted together with the extra Sagittarian energy added by Jupiter through its conjunction with the sun. The contents of the fields of meaning of the dominant archetypes in a chart will characterise the personality and life experiences of its owner, sometimes boosting each other and sometimes cancelling each other out.

When interpreting the planets, tune in to the qualities and themes each planet represents by contemplating its larger archetypal field of meaning. The sun, for example, will carry

Archetypes	Signs	Houses	Aspects
Sagittarius	Sun, Jupiter, Pluto		Jupiter conjunct Sun
Cancer	Ascendant, Moon	Mercury House 4	
Leo	Mars, Neptune	Sun, Jupiter, Pluto House 5	
Pisces	Uranus	Moon House 12	Neptune conjunct Mars
Scorpio	Mercury	Uranus House 8	

meanings from the Leo field into all twelve areas of a chart in the course of a year as it progresses from sign to sign, whereby its Leo qualities will be more enhanced in some signs than in others. For example in Aries and Sagittarius they will be supported by the energies of these extraverted fire signs, and thus shine forth more brightly and creatively. Accordingly we find Blake's sun in a prime position in Sagittarius and House 5, the sun's own house, and further boosted by a conjunction with fiery Jupiter – hence his amazing creativity.

Continuing in this vein we next imagine how the subject matter introduced by a planet combines with the qualities and themes arising through the combination of the archetypal fields of its sign and house. For example Blake's Venus will carry themes from the Taurus and Libra archetypal fields into a segment of the chart characterised by a combination of Capricorn and Virgo characteristics (see Appendix B). To explore the meanings of a planetary placement allow colours, images, ideas and sensations to arise freely in your mind, and above all give the archetypes a chance to inspire you!

Figure 31 shows simplified versions of the Leo and Sagittarius archetypal fields from Appendix E. Meditate on them, bringing the two fields together in your mind and allowing their contents to interact. Create an image for yourself of the kind of person their fusion might describe. This is how you move into a deeper understanding of Blake's sun representing his male side and

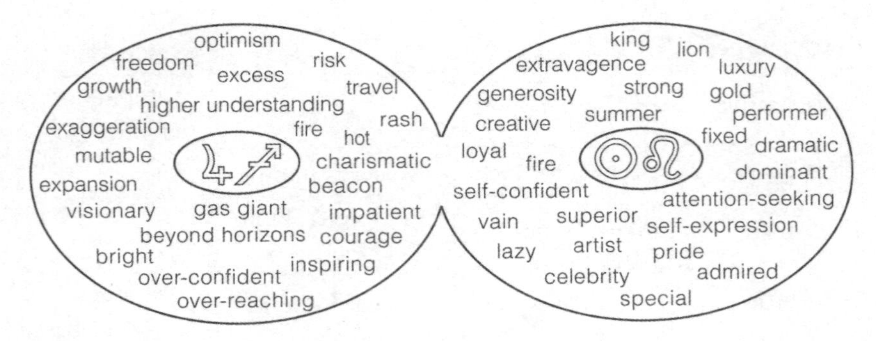

Figure 31. The Sagittarius and Leo fields. Sun in Sagittarius, House 5.

sense of self. Then look at Blake's Sun-Jupiter conjunction in Sagittarius in House 5, and you will understand how the qualities of House 5 (Leo) merge with the aspiring Sagittarian energy to engender Blake's strong creative thrust.

Next we'll explore Blake's moon standing for his female side. You'll need to combine the Cancer (sign) and the Pisces (House 12) archetypes to experience the dynamic relationship between these two archetypal fields created by this combination (see Figure 32).

Eternal wrath and eternal pity

Of all the twelve zodiac archetypes perhaps no two are more unlike than Sagittarius and Cancer – Blake's sun and moon signs. The conflict between these contrary ways of being within himself not only symbolises the contradictions in Blake's personality, but also motivated his attempt to balance the metaphysical opposites in his philosophy.

Blake would make different impressions on different people depending on which side of himself he was expressing. Thus his Victorian biographer Alexander Gilchrist wrote, 'He was energy itself and shed around him a kindling influence,'[12] picking up on Blake's solar Sagittarius side; whereas he was described by a contemporary as 'pale with an expression of great sweetness but bordering on weakness'. This man must have met Blake when

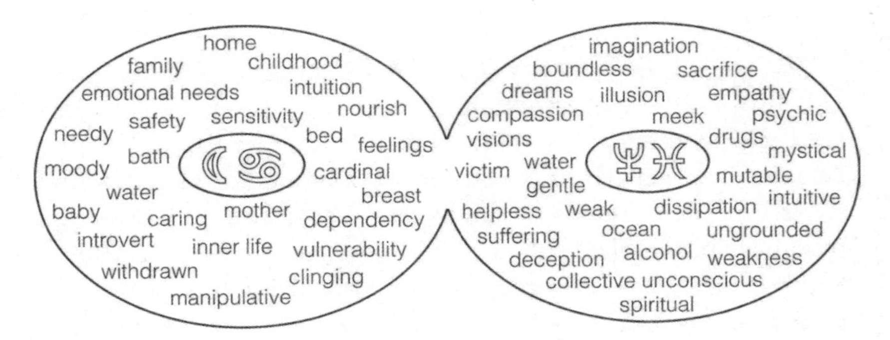

Figure 32. The Cancer and Pisces fields. Moon in Cancer, House 12.

he was in his moon mode. While expressing his fiery side Blake was vigorous and impetuous, and approached people with a bold, fierce directness, whereas when his watery side came to the fore he'd be just the opposite – withdrawn, nervous and hypersensitive.

His painting *Glad Day,* subtitled 'Albion arose from where he laboured at the mill with slaves', expresses his Sagittarian aspiration to live passionately, fearlessly and totally.[13] This side of him breaks out in his *Proverbs of Hell* where he exhorts the reader to cast aside morality and live life to the full. 'Exuberance is beauty', he writes, giving voice to his Jupiter-sun conjunction in fire. 'The road of excess leads to the palace of wisdom' and 'Energy is eternal delight'.[14] Finally, the hymn *Jerusalem,* for which Blake wrote the words, invokes the vital fiery force that fuels artistic creation:

> Bring me my bow of burning gold: bring me my arrows of desire.
> Bring me my spear: O clouds unfold! Bring me my chariot of fire.
> I shall not cease from mental fight nor shall my sword sleep in my hand,
> Till we have built Jerusalem in England's green and pleasant land.[15]

We can see these lines as describing the struggle of the artist to give birth to a work of art, but on a deeper symbolic level (and there are always further depths in Blake) they call for the redemption of Albion, the generic Briton, debased to labouring at the mill with slaves. In this context the building of the ideal city, Jerusalem, becomes an emblem of national and human regeneration. Representing art and architecture proportioned according to the ancient canon (see p. 44), its divine symmetry would then be reflected in a moral, social and economic balance in the land.

In contrast, the people who tuned in to Blake's Cancerian side experienced him as docile, kind and caring. However he could also be very touchy. Sensitive to the slightest slight, he would sulk and hide away when his feelings were hurt. In line with this side of himself, Blake sought security in his domestic routine, and valued his home as a safe place to retire to.

Blake needed longer periods of withdrawal to explore the riches of his inner world and to balance himself emotionally. With his moon in House 12 (the collective unconscious) he would involuntarily pick up on undercurrents of collective feeling, and these, as people who have twelfth-house planets know, can be dark, heavy and emotionally unsettling.

His struggle to reconcile these opposites within himself is the story of Blake's life. And, as it's impossible to be an extravert and an introvert at the same time, he landed himself in dilemmas. For example he would keep on sticking his head provocatively above the parapet, only to be mortally wounded when it was shot at!

In his mythologising imagination the opposite sides of his personality became the two contrary states of the human soul that he called Eternal Wrath, and Eternal Pity. From the former he developed a whole mythological realm called Eden, inhabited by the lions, tigers and wolves howling for prey that we find in the *Songs of Experience,* which for Blake represented the fiery aspects of divinity:

> In what distant deeps or skies burnt the fire of thine eyes?
> On what wings dare he aspire? What the hand dare seize
> the fire? …

> Tiger, tiger burning bright in the forests of the night,
> What immortal hand or eye dare frame thy fearful symmetry?[16]

The might of the tiger symbolises for Blake the creative, or procreative, force of the universe, ever available to those bold enough to seize it. In contrast the lamb and the child are Blake's icons of the state of innocence. In the pastoral landscape of the contrary mythological realm he called Beulah, shepherds guard their sheep, children are loved and cared for and gentleness and compassion reign as in his poem *On Another's Sorrow:*

> Can I see another's woe, and not be in sorrow too?
> Can I see another's grief and not seek for kind relief …
> Can a mother sit and hear an infant groan, an infant fear?
> No no! never can it be! Never never can it be![17]

In Blake's eyes pity was the noblest of female virtues. When he met his wife Catherine he'd just been jilted by another girlfriend, which, based on what we know of Blake, must have sorely dented his pride. He told her all about it and asked, 'Do you pity me?' When she said, 'Yes', he replied, 'Then I love you', and so they got married.

The king and the priest must be tied in a tether

Blake's moon should be seen within the larger archetypal complex created by the triangle connecting it to Uranus and Venus, as the fields of meaning of these two planets will further expand our interpretation of it. We'll start by exploring the psychic terrain of Blake's Uranus in Pisces and House 8 as depicted in Figure 33.

If we combine ideas from the three fields in this figure such as, for example, 'social conscience' (Aquarius), 'compassion' (Pisces) and 'power struggle' (Scorpio), their fusing suggests why Blake was attracted to humanitarian causes, and why he sided with the victims of oppression. We should also note that his Saturn is

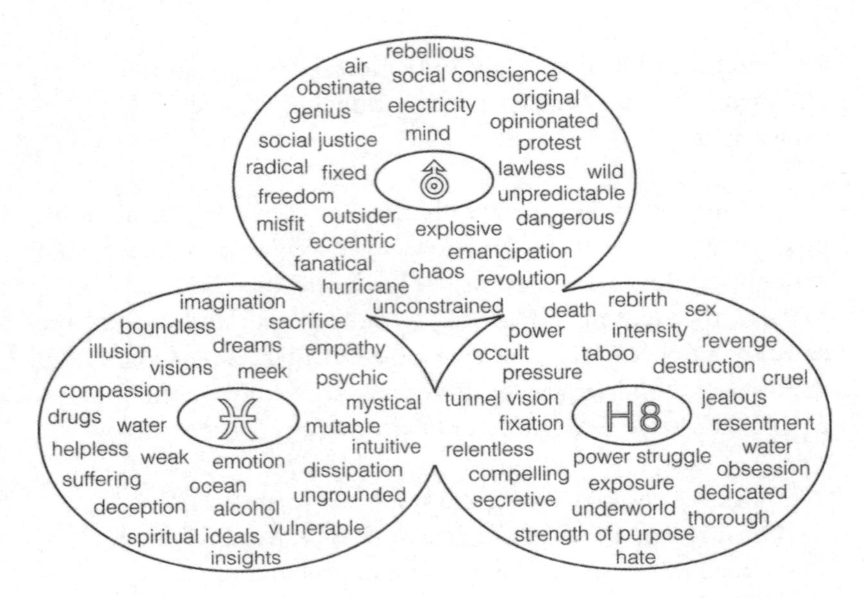

Figure 33. Uranus in Pisces, House 8. The archetypal complex.

in Aquarius in House 7, and as Uranus rules this sign, Saturn's themes and associations play a role in explaining Blake's hatred of tyranny in all its forms. We will see this later in more detail when we look at the square aspect between his Uranus and Pluto.

Blake gives vent to these feelings in his poem *London,* in which he rails against church and king, and blames the institution of marriage for the spread of syphilis.

In every cry of every man,
In every infant's cry of fear,
In every voice, in every ban,
The mind-forged manacles I hear.

How the chimney-sweeper's cry
Every blackening church appalls;
And the hapless soldier's sigh
Runs in blood down palace walls.

But most through midnight streets I hear
How the youthful harlot's curse
Blasts the newborn infant's tear,
And blights with plagues the marriage hearse.[18]

Long before the slogan 'make love not war,' was invented, Blake recognised the psychological link between sexual repression and warfare. He blamed the monarch for the blood being shed on the battlefields, and Christian morality for the disgrace of London's child prostitutes. However his understanding of liberation from oppression was ultimately a metaphysical one. Blake was asserting that modern man will never be free until he has cast off the 'mind-forged manacles' that are enslaving him.

Unusually for a man with a Cancer moon, Blake seems to have distanced himself from his mother and family, as there are no references to them in the records except for a little rhyme in one of his letters revealing that his wife and sister didn't get on. 'Must my wife live in my sister's bane, Or my sister survive on my love's pain?' he wrote – lines that point to an eruption of the tensions in his domestic life that are symbolised in his chart by the Venus-moon opposition.[19] It's likely that he distanced himself from his family when he married, which would correlate with the Uranus aspects to his female planets that symbolise a striving for emotional independence.

To explore the possible ways that themes from the different archetypal fields within a complex could combine, we can take concepts from separate fields and link them up. Thus, if we connect 'marriage' (Venus, Figure 34) with 'freedom' and 'sex' (Aquarius and Scorpio, Figure 33) we can locate Blake's enthusiasm for sexual emancipation in his chart.

Protest against sexual repression runs as a leitmotif through his *Songs of Experience* along with the idea of sexual liberation as a redemptive act. 'The king and the priest must be tied in a tether before two virgins can meet together', and 'The harvest shall flourish in wintry weather when two virginities meet together', he wrote.[20] However, although there are passages in his writings that

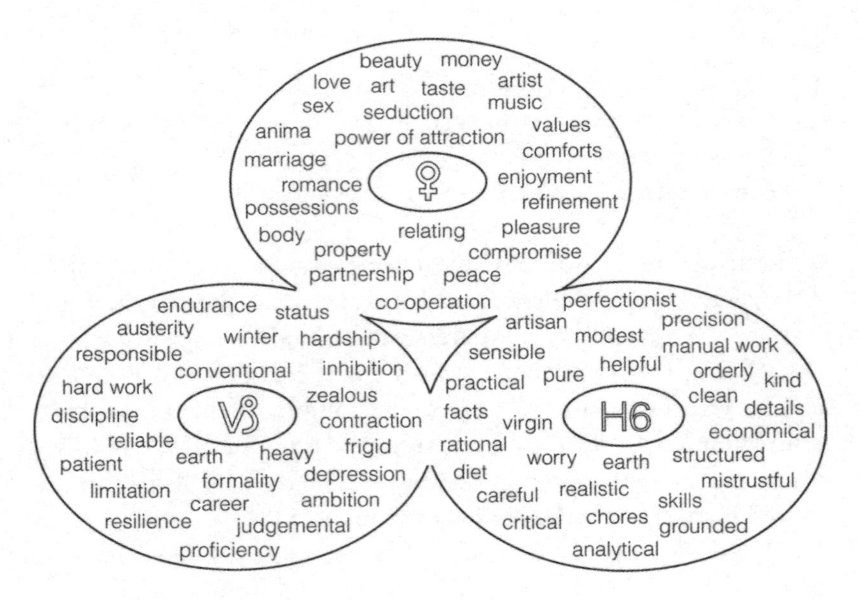

Figure 34. Venus in Capricorn, House 6. The archetypal complex.

seem to advocate unrestrained licentiousness, Blake revered sexual energy as sacred.

So did he walk his talk? Well, in his biography of Blake, Ackroyd mentions contacts in his younger days with groups that experimented with sexual magic and techniques we know as 'tantric'.[21] Also, although he was happily married to Catherine, he was very uncomplimentary about marriage as an institution, as we've seen in *London* where the term 'the marriage-hearse' is coined. But, before we decide whether the Blakes opened up their marriage to third parties or not, we'll first have to interpret Blake's Venus.

The multivalency of the archetypes means that their themes will not always manifest in a birth chart's owner; they can be expressed by other people in his life. Venus in a man's chart is understood in psychological astrology to symbolise one aspect of what Jungian psychologists call the 'anima', and as such is likely to describe his wife or girlfriend. Thus, due to the placement of his Venus, Blake would have been attracted to women with a mixture of Capricorn and Virgo qualities.

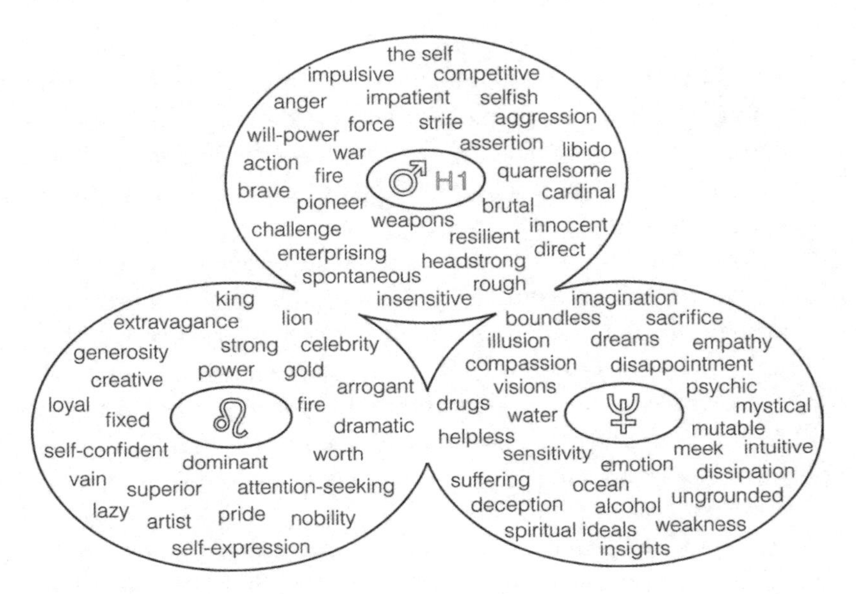

Figure 35. Mars-Neptune conjunction in Leo, House 1. The archetypal complex.

Through merging the archetypal fields illustrated in Figure 34, we can discover a few things about Catherine Blake. For example respectability and social status were important to her as well as material security. Reflecting the earth element (her sun was actually in Taurus) she would have counterbalanced Blake's wilder ideas with doses of healthy realism, and remained no doubt a calm and steadying influence on him. Also we know from anecdotes about her that she was an economical housekeeper and a hard worker. Hayley writes: 'They have no servant: – the good woman not only does all the work of the house, but she even makes the greatest part of her husbands dress, & assists him in his art.'[22]

Another structure in Blake's chart that provides a wider context of meaning for the moon and Venus is the long triangle or *yod* (see Glossary) which links Uranus and Venus to the Mars-Neptune conjunction. As Mars is placed at the apex of this structure, and is inseparable from Neptune, we'll begin with its archetypal complex (see Figure 35).

What strikes us here is that the qualities of Mars as the essence of maleness are nicely boosted by Leo, a masculine fire sign, and by the expansive, extravert ambience of House 1. This Mars could act like a rocket launch pad, providing the thrust for Blake's creative endeavours. Mars also represents the libido, which Blake personified in his epic poems as the giant Luvah, a personification of male virility on a cosmic scale.

Planets in House 1 in a chart feature large in the self-image of its subject; thus we can infer that Blake would have identified imaginatively with Luvah. However Neptune (illusion, disappointment) is part of the equation here, and in House1 can also correlate with self-idealisation. Thus virile posturing on Blake's part could create an illusion leading to disappointment. Also the dramatic way Blake would speak of his genius is reflected by his first-house Leo planets.

Mars and Neptune in conjunction are always uncomfortable bedfellows. Neptune's water fizzles the fire of Mars into steam and mists everything up. Then clarity can become confusion, potency impotency, confidence embarrassment. And, if we put this conjunction in the context of the yod that links it to Uranus and Venus, we can imagine Blake's energy fluctuating between the extremes of wound-up hyperactivity (Uranus overheats Mars) and passive let-go (when Neptune moves in and there is a collapse).

We now understand that his rebellion against sexual constraints (Uranus in aspect to Mars) would shatter like a wave on the hard rock of his Venus in Capricorn. No-nonsense Catherine, who always worried about what other people might think, would have none of that sort of thing. And, as I believe that Catherine held the reins of power in their relationship, I'd put my money on Blake staying faithful to her in deed if not in thought and word!

Mind-forged manacles and dark Satanic mills

We'll look next at a major structure in Blake's chart that provides an overarching context of meaning for six of his planets. It's the fixed *T-square* (see Glossary) created by Saturn in Aquarius opposing Mars and Neptune in Leo and squaring Mercury in

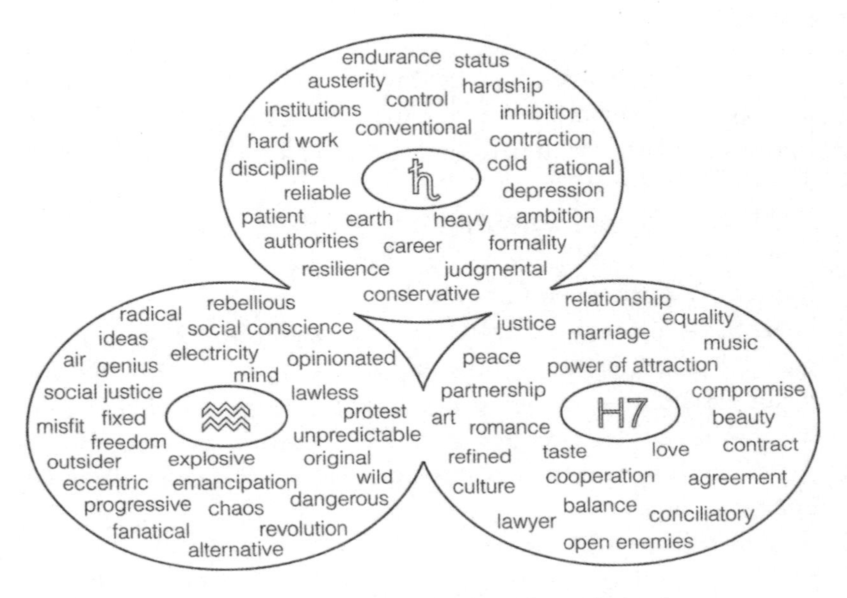

Figure 36. Saturn in Aquarius, House 7. The archetypal complex.

Scorpio in conjunction with the sun and Jupiter (see Figure 36). And as Saturn is the pivotal planet in this complex, that's where we'll start.

Blake painted two great pictures inspired by his Saturn – *The Ancient of Days* and *Newton.*[23] In the latter Newton sits isolated on a rock amidst a barren landscape of rocks and sand. His naked form is beautiful but he's not in his body. He's away in his mind as he measures the arc of a triangle with a pair of compasses. Similarly in *The Ancient of Days* the Almighty is drawing the arcs of time and space with compasses extending from his fingers.

In Blake's private mythology the figure of Saturn looms as Urizen (your reason), the cruel demiurge who created our fallen space-time world. Blake blames this potentate for curtailing our vision, which originally was fourfold, and forcing us to perceive things through the narrow slit of the rational mind. He also blames Urizen's moral laws for the restriction of our natural capacity for sensual delight.

The combination of the Capricorn and Aquarius archetypes inherent in Blake's Saturn (Figure 36) produces a conflict of interests. Progressive and reactionary tendencies clash in this archetypal combination where Saturn's rationality is invaded by the forces of the irrational, and Uranus' fertile chaos threatens Saturn's law and order.

Blake, of course, was identified with the Uranian side of this archetypal mix, and projected the Saturnian on to his opponents (House 7, open enemies). He focused on Bacon, Locke and Newton as the culprits responsible for establishing the scientific materialist paradigm and thus impoverishing our vision and our quality of life. In place of the vibrant living forms that Blake could see shining through matter, all these scientists saw were lifeless atoms, and an atom, as Blake rightly said, is 'a thing that does not exist.'[24]

Saturn opposing his Mars from House 7 points to Blake's main problem – getting on with other people. As an engraver he had to please his customers and keep on the right side of his patrons in order to earn a living, which would be challenging to say the least for a blunt Sagittarius. 'I have no time for seeming and little arts of compliment', Blake once complained.[25]

Saturn manifested in his life as other people who always seemed to be there blocking his path. When he fought them he made enemies. When he withdrew he lost his friends, which led to him becoming a social outcast. The following rhyme, scribbled in one of his many notebooks, expresses the pain of feeling like an outsider:

> O why was I born with a different face?
> Why was I not born like the rest of my race?
> When I look each one starts! When I speak, I offend;
> Then I'm silent and passive and lose every friend.[26]

His sensitivity to criticism didn't help matters. Normally a Mars in Leo in House 1 indicates unshakeable self-confidence. However Blake's self-belief was only skin-deep, as his first-house Neptune reveals. Underneath the bluster he often felt shaky and then would over-compensate. The way he reacted to rebuffs was to become more bombastic, and to assert the truth of his ideas with even greater insistence.

Neptune is always accompanied by a certain nebulousness, and Blake, who was unable to see himself clearly because of it, tended to either over-estimate or under-estimate his achievements. Failures triggered his self-doubt, and doubt for Blake was the most fatal of flaws. 'If the sun and moon should doubt, they'd immediately go out!' is one of his well-known sayings.[27]

The clash of the Aries (Mars) and the Capricorn (Saturn) archetypes, as in Blake's square aspect between them, signifies a potential problem with anger, and Blake's violent temper is well documented. Ackroyd quotes reports of the 'unmeasured violence' of Blake's speech when he was provoked by opposition – a reflection of the link-up between Saturn, Mars and Mercury in his chart in that tense T-square.

In line with Saturn in transpersonal Aquarius it could be triggered by social injustice, and by the cases of cruelty he would often witness in the London streets. If we combine the qualities from the Neptune-Pisces and the Mars-Aries archetypal fields (Figure 35), we can understand Blake's anger in the context of his strong empathy with the sufferings of the victims of abuse and oppression.

An exploration of the meaning of his Mercury in the context of the T-square can help to explain why Blake was misunderstood. First we need to make a synthesis of Mercury's archetypal components by combining its planetary characteristics with the qualities of the Scorpio and Cancer fields.

Mercury in a chart symbolises the kind of mind the owner has as well as what he tends most to think about. And Blake's Mercury in Scorpio and House 4 (a water sign and a water house) is the signature of an emotional rather than a rational thinker (see Figure 37). So it's not surprising that Blake expressed his ideas through symbolism and metaphor rather than rational argument, especially as Mercury and Jupiter square Neptune. Also his Mercury's fusion with the Scorpio archetypal field brought subjects like sex, death and the occult to the fore in his mind.

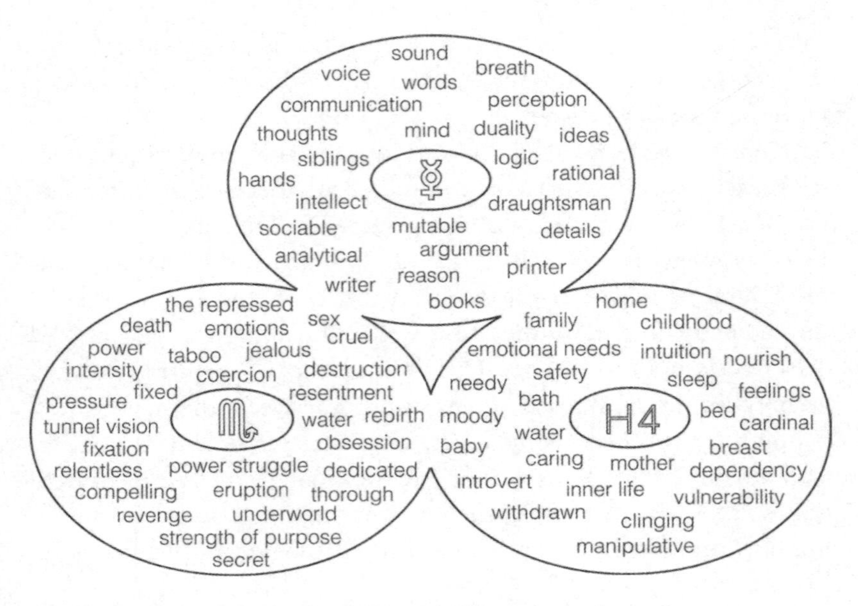

Figure 37. Mercury in Scorpio, House 4. The archetypal complex.

Scorpio's fixity and depth are reflected in the stubbornness of his opinions and also in the intensity with which they were expressed.

The belligerence of Blake's Mercury, exacerbated by its square aspect on one side to Mars, and on the other blocked by its square to Saturn, is reflected in the angry polemic of some of his prose, and in the rude annotations he was want to scribble in the margins of his books when their authors had said things that annoyed him.

To a large extent, Blake was himself to blame for his social isolation. When invited out to dinner as a 'curiosity' he would alarm the other guests by describing the disembodied spirits he saw hovering about them in the room. On other occasions, under the influence of his rebellious Uranus, he'd shock the more conventionally-minded company present by making provocative remarks.

Even John Linnell, Blake's patron and main support in his last years, was embarrassed. 'With all the admiration possible', he wrote:

> ... it must be confessed that he said many things tending to the corruption of Christian morals, even when unprovoked by controversy, and when opposed by the superstitious, the crafty, or the proud, he outraged all common sense and rationality by the opinions he advanced.[28]

All we see is vision

Blake was misunderstood because his key ideas were too far ahead of his times. It would take till the 1960s for a generation to arrive that was open to them. Blake must have foreseen this as he wrote in *The Songs of Experience:*

> Children of the future age
> Reading this indignant page,
> Know that in a former time
> Love! sweet love! was thought a crime.[29]

And the Blake enthusiasts amongst the baby-boomers, liberated by the pill and experimenting with a sexual freedom that Blake could only dream of, thought he was speaking to them!

It was also during the sixties, another decade marked by a powerful conjunction between Uranus and Pluto, that Blake's work began to be appreciated not only amongst hippies on the road but also in the universities. Revolution was a buzz word again, and his radical metaphysical, aesthetic and sexual values had suddenly become interesting.

Tarnas' research into past Uranus-Pluto alignments has shown that they coincide with times of radical social and political upheaval. Rebellions against all forms of oppression break out, and wars and revolutions erupt as the entrenched powers within the state attempt to resist the onslaughts of radical change.[30] However these are also periods of amazing artistic and technological creativity, and of quantum leaps in human consciousness. Try meditating on the effects of a Uranus-Pluto synthesis by merging together the Aquarius and Pluto (Uranus and H.8) archetypal fields from Figure 33 in your mind's eye.

The present Uranus-Pluto square, prominent during the second decade of the twenty-first century, resonates diachronically with the 1960s as well as with the period of the French Revolution when Blake wrote the lines quoted at the beginning of this section. Once again cracks and fissures are appearing in the social and political structures worldwide, and the molten lava of revolution and war is pushing through. However times of calamity are also times of vision, and when the bottom is threatening to fall out of our world, we become receptive to new paradigms.

Uranus and Pluto work in alliance to purge all that's rotten in the collective mind and overthrow all obsolete systems. And the people with strong Uranus-Pluto configurations in their charts become instruments in furthering this work. Hence the subversive revolutionary message of Blake's writings, which his disciple Frederic Tatum (who inherited the bulk of Blake's papers on Catherine's death) found so threatening that, tragically, he destroyed a large chunk of the material!

Blake's fifth house Pluto, which I weight as especially strong as it stands in the same house and sign as the sun and in a wide conjunction with it, was reflected in his compulsive creativity and obsessions. It can also explain the high pitch of intensity with which he delivered his message. However, for many people Blake was simply too fanatical, and instead of convincing them he put them off.

On the positive side Blake's strongly placed Pluto allowed him to tap into the deeper strata of the collective unconscious from whence he drew inspiration. Pluto marks the spot in a chart where something powerful and archaic can push up from the depths. When aligned with Neptune, as in Blake's case through a trine, it can signify a transformative mythic vision. And I suggest the reason why his most popular poems are so unforgettable is because their roots go right down to the archetypal layer in the collective psyche.

In my view Blake's prophetic books are all about archetypes. *The Four Zoas,* for example, tells of the 'four mighty ones' that are in every man. These titanic cosmic principles are the result of a primordial fall out of oneness whereby, as in Egyptian cosmogony and in the tetraktys, the one became two and then the two became four.

His Zoas remind us of the ancient deities of the four cardinal directions (see p. 117) and like them also correspond to the four elements of earth, water, air and fire. Blake extends their influences into the macrocosm and microcosm in a vast system of fourfold correspondences. Then the fall continues with the four Zoas pairing with their female emanations and fathering twelve principles that Blake calls the Twelve Tribes of Israel. Sounds familiar, doesn't it? So where did Blake get these ideas from?

Well, rather than borrowing them from the sparse literature on world mythology that was available to him, I believe he independently plumbed the collective unconscious and dredged them up from its depths. Jung also had this capacity. And, in line with the way Jung saw the archetypes, Blake's mythological characters are both psychological states (the microcosm) and forces within the cosmic mind (the macrocosm).

As well as describing a fall in consciousness, Blake's writings present a path of potential spiritual regeneration. 'I rest not from my great task to open the eternal worlds, to open the immortal eyes of man inwards, into the worlds of thought, into eternity,' he wrote.[31] It must be his strong Neptune in House 1, boosted by his twelfth-house moon together with his Mercury conjunct Jupiter squaring Neptune, that represents the gift of vision in his chart.

> To see a world in a grain of sand,
> And a heaven in a wild flower,
> Hold infinity in the palm of your hand
> And eternity in an hour.[32]

Blake is saying that we can look at a flower through the spectacles of mental understanding, in which case what we see is an abstraction – our mental image of the flower. Or we can look with the immediacy of first-hand vision and then the flower transports us to heaven. This is more than second sight; what Blake is talking about is 'fourfold vision'.

> Now I a fourfold vision see,
> And a fourfold vision is given to me;
> Tis fourfold in my supreme delight
> And threefold in soft Beulah's night,
> And twofold always. May God us keep
> From single vision and Newton's sleep![33]

Single vision, the consequence of the Newtonian worldview, is like trying to see through eyes that are clouded by cataracts. 'For man has closed himself up till he sees all things through narrow chinks of his cavern,' Blake wrote; whereas 'If the doors of perception were cleansed everything would appear to man as it is, infinite'.[34] Blake passionately believed that the way forward was through a revolution in perception, which would give rise to a new paradigm of reality.

So where do we find this experience of spiritual expansion in his chart? Perhaps symbolised by the conjunction between Mercury (perception) and Jupiter (expansion), squaring Neptune (vision). This structure, however, had a downside for Blake as a writer, correlating as it does with boundlessness and the loss of structure. His verse in many passages in his prophetic books is like flood water overflowing the bounds of form. Literary critics consider these books to be marred by too many sections that are rambling and repetitive.

However Blake didn't consider it necessary to use the regulating intellect to prune what he wrote. He trusted implicitly in his inspiration (or in the voices he heard in his head dictating to him!), and he over-taxes his readers as a consequence. This weakness of style, together with his iconoclastic use of symbolism explains why his epic poems have been called the 'wild effusions of a distempered brain'.

Yet they also contain lines of pure, stark beauty: 'The vegetable universe opens like a flower from earth's centre in which is eternity. It expands in stars to the mundane shell and there it meets eternity again both within and without.'[35] What about that?

Dare frame Thy Fearful Symmetry

As we have seen from our example interpretation, the signs, houses and planets in a chart symbolise different facets of the twelve astrological archetypes. In the geometry of their inter-relationships lies the source of the qualities that emerge in the subject's personality, and in the themes prominent in his life story. Thus astrology with its overarching framework puts our experience of our self and of our life history into a cosmological context, and in doing so makes it meaningful.

Le Grice writes:

> Archetypal astrology, by making possible the recognition of just these hidden archetypal principles and themes, might restore to each individual life a sense of meaning, derived nor from any religion or doctrine, nor from any philosophy or creed, but from our own personal relationship to the archetypal powers of the cosmos.[36]

Astrology reveals the mythic perspective of our lives. The patterning of the archetypes in our charts plays out in narratives with mythical depth and universal significance. And, because each life is a unique realisation of the potential of the archetypal structures, although the same stories play out again and again each has a unique inflexion. The multivalency of the archetypes make a multitude of variations on the small number of basic myths possible, and these are manifesting in your life, in my life and in everyone's lives right now.

Blake's myth is that of a highly gifted artist and poet who had an important message to deliver to his generation, which, because of certain faults in his character, was not heard by his contemporaries. And, in spite of all his efforts, his achievements in art and poetry were not acknowledged. It echoes the myth of Tiresias, Homer's blind prophet, who was spurned by society and cast out into the wilderness.

However, although Blake lived in obscurity during his lifetime, his genius is recognised today, and he has his rightful place as a seminal figure in the poetry and painting of the Romantic period. In 2002 he was even voted number thirty-eight in a BBC poll of the hundred greatest Britons!

And when at the opening ceremony of the 2012 Summer Olympic Games in London the crowds packing the Olympic stadium joined with the choir to sing *Jerusalem*, and the words of Blake's poem rang out not only across London, his hometown, but across the whole wide world (there were an estimated 900 million television viewers that night!), poor, neglected William Blake had his long delayed moment of glory. What a turn up for his books!

Conclusion

But perhaps there is a pattern set up in the heavens for one who desires to see it and, having seen it, to found one in himself. (Plato)[1]

Ancient monuments the world over were designed in accordance with the same scheme of proportion, and used the same units of measurement, the reason being that these were derived from celestial patterning. The lucid, moving geometry of the sky, created by the mathematical inter-relationships between the heavenly bodies, served the ancients as a model for perfection. They emulated the harmony they observed in the sky in their social order, aspired to in their ethics, and mirrored it in the proportions of their art and architecture.

How urgently we need such a value-giving cosmology in our present age of unprecedented challenge! As Le Grice expresses it in *The Archetypal Cosmos:*

> We are living with the decline and dissolution of the old order, with the tumult and uncertainty of a new birth. The need for an orienting context to guide us through these many changes has become urgent.[2]

Prophetic thinkers, such as systems scientist Ervin Laszlo, are warning of an imminent global breakdown. As he wrote in 2006: 'In the next few years, new thinking and new action will be crucial; without them our globalised systems could break down in chaos'.[3]

This was not news to astrologers either then or now. They've long been aware of the drastic planetary transits marking the period between 2008 and 2025, which include a 'make or break' square aspect between Uranus and Pluto with an exceptional seven *exactitudes* (see Glossary). This is not business as usual! As Laszlo points out, we have reached a tipping point in our planet's history at which we are confronting a division of the ways; one leads towards global collapse and the other to reorientation and renewal.

In my view the real need of our times is for a paradigm switch. The worldview of scientific materialism, that has led to us seeing the world as through the narrow chinks of a cavern, needs to be expanded into a more life-enhancing cosmology. The Jungian analyst Anne Baring in her book *The Dream of the Cosmos: a Quest for the Soul* writes of our present challenge:

> What is the emerging vision of our time which could offer a template for a conscious, awakened humanity? ... The answers we seek cannot come from the limited consciousness which now rules the world but could grow from a deeper understanding, born of the union of mind and soul, helping us to see that all life is one; that each of us participates in the life of a cosmic entity of immeasurable dimensions.[4]

The emerging archetypal cosmology could be just what's needed. It affirms that oneness and interconnectedness are a deeper truth than the separateness of things, and echoes the ancient vision of the universe as a plenum or unified whole, with its components all interconnected in a web. This is the universe that Plotinus describes in the *Enneads:*

> This one universe is all bound together in shared experience and is like one living creature, and that which is far is really near ... and nothing is so distant in space that it is not close enough to the nature of the one living thing to share its experience.[5]

He was looking at the world through a very different lens to that of our modern-day materialist-atheists; and what he saw through it was human life embedded in an alive and conscious cosmos!

There are some hopeful signs that the archaic vision of oneness could be in the process of re-emerging in human collective consciousness. The conception of our planet as the goddess Gaia is gaining ground and bringing a shift from exploitation to guardianship in our attitude towards the natural world. Some cutting-edge thinkers are expanding this conception to apply to the whole universe, seeing it is an alive, intelligent organism.[6] And if this cosmological revolution takes hold during the twenty-first century it could be our planet's saving grace.

The implication is that, if the universe is essentially one organism, conscious, self-regulating and self-determining, then human individuals are one with each other and with all creation. Bodies are the only things that are separate – nothing else! And if this conception of our fundamental unity with all things becomes our lived experience, our compassion will increase. We'll no longer believe we are in competition with others. And we'll hesitate before we harm others, because this would mean we are harming ourselves (which is already the case, though most people are unaware of it).

The Bible says that man was made in the image of God. However the fuller truth may be that the cosmos is made in the image of man. Then, like a human being, the universe can be seen as having a body (its material parts), a mind (its intelligence), and a psyche with an inner life (the *anima mundi).* The theologian Origen, who lived in the third century AD, saw it this way when he wrote: 'Understand that you are another little world and have within you the sun, the moon and also the stars.'[7]

That the universe has intelligence is proved by its brilliant mathematics and the clever way it self-regulates. And astrology suggests that meaning in the cosmos derives from the universal mind as the source of the all-encompassing mathematical and geometric basis of creation. That the universe has a feeling soul is demonstrated by the exquisite beauty of the colours found in flowers or in the markings on birds, which reveal refined aesthetic sensitivity and perfect taste.

The enlightened Greek philosopher Heraclitus spoke about the *anima mundi* when he said there never was a time when this universe did not have a soul, or the body existed in the absence of soul, and that '*You could not discover the limits of the soul, even if you travelled by every path in order to do so; so profound is its meaning*'.[8]

If, as this book has suggested, the universal mind and psyche are present and active throughout the universe, and each human individual participates in them, then the same archetypal patterns within the psyche of the whole that determine the quality of events arising in the outer world, also determine the events and quality of individual experience. The two realms of above and below, are really one, both deriving from the same archetypal principles. And, as I have shown, an adequate explanation of how astrology works is provided by the implications our 'separate' minds being fractals of the one universal mind, and so imprinted with its archetypal patterning.

When macrocosm and microcosm are seen as two planes on which the intentions of the universe manifest in parallel, the phenomenon of synchronicity can be explained. And, as the synchronicities registered by astrologers between events in the sky and events on earth can be scientifically proved, the aura of charlatanry and the dubious arts, which has surrounded astrologers like a dark cloud for many centuries, is dispelled.[9]

We have seen how astrology rests on the truth of the big secret of 'as above so below' and that this should be understood quite literally. We've also seen the extent to which meaning in the cosmos is structured by the twelve astrological archetypes. In fact any sense we can make of human life derives from them. And that these archetypes are central ideas in the cosmic mind, acting as nodes of meaning around which fields of associated ideas cluster.

Also the model of the great chain of being with its cross-correspondences between the realms can be used to understand how the subsidiary ideas within these clusters are categorised within the universal mind. It's also an apt metaphor to describe the structure of the qualitative dimension of the universe that is drawn on by the symbol system of astrology.

If an astrologer has developed what Jung used to call the 'archetypal imagination', she will be able to intuitively tap into the archetypal fields of meaning in the cosmic mind. However a training in meditation is needed for this in my experience.[10] Astrologers need the capacity to go inside and concentrate on the inner. They will need to move down through their personal thoughts and feelings that are illustrated by the top layer of Jung's iceberg (see Figure 2, p. 23) to access the deeper levels of the collective mind where the archetypes dwell. Then interpreting meaning becomes an act of co-creativity.

Astrology demonstrates how the archetypes relate geometrically to one another within great interlocking time cycles to create universal patterns of meaning. Using an *ephemeris* (see Glossary), we can observe how individual archetypes rise into prominence within these cycles, and then recede into the background again. A familiarity with the contents of their archetypal fields of meaning will allow us to recognise their signatures when they have emerged. Each day, indeed each hour and minute, is characterised by a specific archetypal mix, which colours what we do, feel and think, affecting not just our personal lives but also the course of history.

The archetypes within the unending creative process that is universal consciousness combine in our personalities, and manifest in our lives, to give human experience a mythic perspective. Thus our little life, which appears so insignificant in the context of the millions of planets, stars and galaxies, expresses a unique combination of the twelve archetypal forces that order the cosmos. Each one of us, therefore, is making a unique contribution to the larger cosmic patterns of meaning being woven by the whole.

Astrology was revered as the king of sciences in the Egyptian temples because it bears witness to a coherent and meaningful universe. And the world today urgently needs this proof that what happens is not random, which is why Tarnas' book *Cosmos and Psyche,* which offers convincing evidence that planetary cycles and world events are correlated, is one of the most positively paradigm-changing works of our time.

By demonstrating that meaning is intrinsic to the workings of the cosmos, astrology has the power to transform our generation's paradigm. It can awaken us from the belief that the universe is random and meaningless. And when we have become aware of the astrological archetypes manifesting in and around us, we live our lives more consciously as a consequence

The conception of an open-ended universe, one that is not hide-bound by immutable laws, is intrinsic to the new emerging cosmology and in line with the postmodern scientific paradigm. There are scientists today who no longer see the laws of nature as eternal and unchangeable, and this must necessarily be the case in a universe that is alive and conscious. If the universe has a mind, it must have the power to change it!

So the archetypes are eternal, dwelling as they do outside time, but their expression in the context of a universe in flux will always vary. For this reason astrological pronouncements should not be understood in a deterministic way. Tarnas coined the phrase 'archetypally predictive' to describe the blend of continuity and innovative freedom characterising how the astrological archetypes manifest. And it's because of their flexibility that outcomes can be influenced by human intervention, which the ancients knew and hence their custom of making offerings to the gods, while prayers are used for this purpose today.

The proven fact that we create the world through the act of perceiving it is a factor that points to our co-creative role in the cosmos. Dennis Elwell in *Cosmic Loom* goes further – 'we are able to appropriate our personal quota of the directional power of the universe', he writes, and the very fact that we have will power and frame intentions, he says, confirms 'the presence of a directional will inherent in the universe itself'.[11]

As an integral part of the universe we are affected by the whole, but we are also continually affecting the whole through how we think, feel and act. In other words, not only do the stars influence us but we in our turn influence the stars – which is a very big idea to take in!

Our voluntary (or involuntary participation) in the direction the universe is taking means we share responsibility for it. Theoretically at least, the course of the universe can be modified by our will and intentions. That is why, for some time now, people have been banding together in meditation groups and prayer circles the world over to use their combined wills to divert the course of human destiny away from the worst of possible outcomes. And that we have so far survived being wiped out by nuclear war, nuclear accidents, climate change induced natural disasters and global pandemics could be, in part at least, due to their efforts.

The time is ripe, especially during this present revolutionary decade, for a paradigm shift in astrology which will give it a new philosophical and methodological basis. In my view archetypal astrology, which provides a framework for cosmically based meaning, and recognises our co-creative role in the unfolding of archetypal potential, could be just what's required. Le Grice announced in 2010 that:

> A new form of astrology – archetypal astrology – has now emerged, drawing on the astrological tradition yet informed by the insights of depth psychology and supported increasingly by the theoretical conceptions of some of our finest scientific minds.[12]

The archetypal approach to astrology he describes is in line with the new cosmology as it approaches astrological meaning 'top-down', both philosophically and on the practical level of chart interpretation. In contrast the philosophical roots of the western astrological tradition are planted in Aristotelian reductionism and the static Ptolemaic cosmos. Its approach to interpretation is 'bottom up' thus reflecting the procedures of reductionist science.

The top-down approach of archetypal astrology reflects the

conception of a universe ordered by an immanent universal intelligence. Single factors within a chart are thus seen as components of larger frameworks of meaning, all ultimately deriving from the patterning of the astrological archetypes within the universal over-mind.

We live in a world today that's very different from the Classical world or the world of the Middle Ages. And the meanings of the traditional astrological significators have changed, while new significators such as Uranus, Neptune, Pluto (and lately Chiron) have emerged as carriers of meaning. I believe that astrology, like the universe itself, should be seen as a process in flux. The parameters it is based on can change, and its pronouncements must be variable. If they are understood deterministically, astrology loses touch with the dimension of the sacred and becomes a mere system of divination.

The paradigm of an archetypal cosmos could lead to a new form of spirituality in which astrology plays a central role, providing us with reassuring proof of the presence of a transcendent order. It would encourage an increase in sensitivity to the spiritual dimension. We may begin to see 'a world in a grain of sand and a heaven in a wild flower', which would lead to an increased reverence for the earth and for all life upon it.

As in ancient times, the astrological archetypes would be seen as divinities again, and the closer we came to them the more we would be open to transforming experiences of love, truth and beauty. Archetypal astrology, with its power to unlock a portal to the sacred, would be an integral part of the alchemical path of spiritual refinement, as in the Egyptian academies and the monasteries of the Pythagorean brotherhood. Ptolemy penned the following epigram in this vein:

> I know that I was mortal born, creature of but a day. But when the revolving spirals of the stars in my mind I trace, my feet touch earth no more. I feast with Zeus, filled with ambrosia, nourished as the gods themselves.[13]

Meditating on the night sky was for the ancients a spiritually elevating experience in which their consciousness became aligned with 'the divinities that move the stars'. A devotional approach to the heavenly bodies can lead to a melding with the ineffable archetypes manifesting through them. Try it for yourself. Meditate for example on Venus as morning star or on Sirius and you may experience that heightened state of consciousness described by Ptolemy, accompanied by uplifting feelings of awe, veneration, and love, because the stars, as Plato expressed it, are 'the great visible gods'.

The climax of the initiation of the neophytes in the mystery schools of Egypt and Greece was attained, so we are told, when they watched the wheels of cosmic motion turning, and experienced the rhythms and harmonies by which the universe is maintained within the unfolding geometry of perpetual creation. The writer of the apocryphal *Book of Enoch* experienced such a vision, and it's with his account that I will leave you:

> And I saw the holy sons of God. They were stepping on flames of fire: Their garments were white and their faces shone like snow.
>
> And I saw two streams of fire. And the light of that fire shone like hyacinth. And I fell on my face before the Lord of Spirits.
>
> And the angel Michael seized me by my right hand, and lifted me up …
>
> And he showed me all the secrets of the ends of the heaven, and all the chambers of all the stars, and all the luminaries, whence they proceed before the face of the holy ones.
>
> And he translated my spirit into the heaven of heavens …[14]

Glossary of Astrological Terms

Angles: the *ascendant, descendant, medium coeli (MC) and imum coeli (IC).*

Angular: term used in traditional astrology for the cardinal houses 1, 4, 7 and10.

Ascendant: the point in a chart symbolising the place where the ecliptic cuts the eastern horizon.

Aspect: the geometric relationship between two planets.

Cadent: the term used in traditional astrology for the mutable houses 3, 6, 9 and12.

Cardinal: the first of the three modes of energy, see chapter 6.

Chart: a horoscope diagram

Conjunction: the relationship between two or more planets connected through proximity and lying within a predetermined *orb,* see Appendix D.

Cusp: the beginning of a house at the dividing line between it and the previous house.

Decan: A ten-degree section of the ecliptic. Each astrological sign covering thirty degrees of the zodiac is divided into three decans further differentiating the signs' meanings.

Descendant: The point in a chart symbolising where the ecliptic cuts the western horizon

Ecliptic: The circle drawn by the sun against the background of stars in the course of a year, as seen from the earth. Its division into twelve equal sections creates the signs of the *tropical zodiac.*

Element: the four elements used in astrology are fire, air, water, earth, see chapter 6.

Ephemeris: a book of tables showing among other things the positions of the sun, moon and planets over a set time period, for example 50 years.

Equal houses: a system of dividing the inner wheel of the houses into twelve sections each measuring thirty degrees, see Endnote chapter 7, 7.

Exactitude of a *transit*: when an aspect is exact (without any orb), for example a transiting planet is on the same degree and minute of the degree as another planet.

Fixed: the second of the three modes of energy, see chapter 6.

Glyph: sign or planet symbol, see Appendix A.

Heliacal rising: rising at dawn. The adjective 'heliacal' describes stars that appear on the eastern horizon immediately before sunrise.

Houses: the twelve sections of the inner wheel in the horoscope diagram.

Horoscope: an astrological chart (also a horoscope column in a newspaper).

IC/ 'imum coeli': Latin for 'deepest heaven', the lowest declination of heavenly bodies below a local horizon.

MC/ 'medium coeli': Latin for mid heaven, the highest declination of heavenly bodies above a local horizon.

Mode/Modality: the cardinal, fixed and mutable energy forms, see chapter 6.

Mundane astrology: the astrology of countries, towns, groups, institutions, events etc in contrast to the personal astrology of natal charts.

Mutable: the third of the three modes of energy, see chapter 6.

Natal chart: a horoscope drawn up for the time and place of an individual's birth.

Negative signs: the female signs Taurus, Cancer, Virgo, Scorpio, Capricorn and Pisces.

Opposition: the aspect formed between two planets or points in a chart that lie 180° apart, see Appendix D.

Orb: The variance from the geometric exactitude which is astrologically allowed. Some astrologers allow wider orbs than others. Note that different types of aspect are allowed different sized orbs, see Appendix D.

Personal or inner planets: Sun, moon, Mercury, Venus, Mars and possibly Jupiter.

Planets: In astrology the sun and moon are referred to as 'planets'. In modern astrology the planets usually used are sun, moon, Mercury, Venus, Mars, Jupiter, Saturn, Uranus, Neptune, Pluto and increasingly Chiron.

Planetary Cycles: dynamic unfoldings of the twelve archetypes in sequence as represented by the planets.

Positive signs: the male signs Aries, Gemini, Leo, Libra, Sagittarius and Aquarius.

Precession/precessional cycle: the 26,000-year cycle of the spring equinox point through the zodiac, see Endnote chapter 2, 7.

Progressions: Techniques of progressing the planets and sensitive points in a chart to obtain predictions for the future.

Quadrant: the wheel of the houses is divided into four quadrants or quarters by the *angles* that mark the cardinal directions.

Quadruplicities: the three grand crosses, see chapter 6.

Quincunx: the 150° aspect see Appendix D.

Quintile: the 72° aspect. It carries mystical overtones because the significant numbers 72 and 5 are involved (5 x 72 = 360), see Appendix D.

Ruler: The planets traditionally each have a sign that they rule and in which they are especially strong. These signs correspond to them archetypally.

Sextile: the 60° aspect, see Appendix D.

Sidereal zodiac: a zodiac with divisions based on the twelve constellations of stars rather than on the geometric division of the *ecliptic* into twelve sections.

Succedant: the term used in traditional astrology for the fixed houses 2, 5, 8 and11.

Semi-sextile: the 30° aspect, see Appendix D.

Square: the 90° aspect, see Appendix D.

Sun sign: the sign in which the sun is placed in a natal chart.

T-Square: An aspect pattern in the form of a triangle composed of an *opposition* and two *squares*.

Transits: the hour by hour or day by day movements of the sun moon and planets across significant points in a horoscope.

Transpersonal planets/outer planets: Saturn, Uranus, Neptune, Pluto and Chiron.

Trine: the 120° aspect, see Appendix D.

Triplicities: the four grand trines, see chapter 6.

Tropical zodiac: a *zodiac* in which the *ecliptic* is divided into twelve equal sections of thirty degrees.

Unequal houses: see Endnote chapter 7, 7.

Whole sign houses: see Endnote chapter 7, 7.

Yod: An aspect pattern forming a triangle with angles of 60, 150 and 150 degrees.

Zodiac: a belt about 16 degrees wide on either side of the line of the ecliptic within which the orbits of sun, moon and the main planets pass.

Endnotes

Introduction

1. David Peat *Synchronicity: the Bridge between Mind and Matter,* p.103.
2. See for example Graham Hancock & Santa Faiia *Heaven's Mirror: Quest for the lost Civilization.*
3. Richard Tarnas *Cosmos and Psyche: Intimations of a New World View.*
4. Keiron Le Grice *The Archetypal Cosmos: Rediscovering the Gods in Myth, Science and Astrology.*

Chapter 1

1. *Lebensrad* (1886), Burk Verlag, Burgdorf, Germany.
2. Carl Jung *Synchronicity: an Acausal Connecting Principle, Collected Works of C.G. Jung,* Vol 8.
3. See David Bohm *Wholeness and the Implicate Order.*
4. Tarnas *ibid.*
5. David Peat *Synchronicity: the Bridge between Mind and Matter.*
6. Roger Woolger *Other Lives, Other Selves.*
7. Fritjof Capra *Turning Point,* p.17.
8. Glenn Clark *The Man who Tapped the Secrets of the Universe,* p.43.
9. Quoted by John Anthony West & Jan Gerhard Toonder in *A Case for Astrology,* p.267.
10. Quoted by Le Grice *ibid.* p.152.

Chapter 2

1. Nicholas Campion *Astrology and Cosmology in the World's Religions,* p.4.
2. Le Grice *ibid.* p.37.
3. West & Toonder *ibid.* p.26.
4. John Michell *City of Revelation,* p.109.
5. See Hancock & Santa Faiia *ibid.*
6. Herodotus *The Histories* Vol 2, p.186.
7. The word precession means 'backwards' and refers to the movement of the spring equinox point through the 360 degrees of the zodiac at

the rate of one degree in 72 years in reverse direction to that of the sun and planets. This means that, in the Great Year or precessional cycle of 26.000 years, the Age of Pisces is followed by the Age of Aquarius whereas normally Aries follows Pisces. Imagine the earth's axis as a pencil stuck through the earth from its south to its north pole. This pencil is then extended upwards into space to align with the pole star. Due to the gravity of the sun and moon, the earth's axis wobbles as it turns, which causes the point of the imaginary pencil to draw a circle in the sky like that described by a spinning top.

The phenomenon of precession has two important consequences for astrology. Firstly that the zodiac sign in which the sun rises at the spring equinox will change every 2100 or so years marking the shift of the Great Ages from one zodiac sign to the next. At present there's a shift going on from Pisces to Aquarius. Secondly, within the precessional cycle the constellations appear to rise and fall in relation to the horizon. For example the constellation of Orion is at present reaching its highest declination which is meaningful to astrologers.

8. See R.A.Schwaller de Lubicz *Sacred Science* and *The Temple of Man.*
9. See Paul Devereux *Stone Age Sound Tracks: the Acoustic Archaeology of Ancient Sites.*
10. Christopher Knight & Robert Lomas *Uriel's Machine: the Ancient Origins of Science,* p.350.
11. Researchers are recognising that the stone circles and Neolithic structures called tombs were more than burial sites. It's being suggested that they were cult buildings where, in addition to funerary rites, initiatory rituals concerning death and resurrection were celebrated. The use of sonics, and the numbers and geometry involved in their architecture and astronomical alignments, could have been consciously employed to facilitate powerful spiritual experiences.

    These structures, that are microcosms of Neolithic cosmology, reveal something of the beliefs of the people who created them. The numbers that generate their proportions, the geometric figures appearing in their lay-out, the symbols engraved on significant stones, and the ubiquitous symbolism of the four elements – water (proximity to a river, a moat, large basins containing water), fire (hearths), air (excarnation platforms), as well as earth (wood, stone and earth as building materials) – embody cosmological concepts. Above all what they have in common is a conscious mirroring of the changing relationships between sky and earth, as numbers and geometry derived from the rhythms of sky are reflected in their structures.
12. Pyramid Text 508 www.pyramidtextsonline.com, trans. Faulkner, Piankoff, Speleer.
13. Rosemary Clark (2000) *The Sacred Tradition in Ancient Egypt*, Llewellyn, p.193.

14. Michell, *ibid.* p.125.
15. *Stobaei Hermetica* ed. Nock and Festugiere XXIII 2–3.
16. For example, whatever her private views, orthodox Egyptologist Lucia Gahlin in her books *Egyptian Religion* and *Egyption Myth* approaches her material from the materialist, scientific understanding that Egyptian gods were collective projections, and Egyptian religion was erroneous superstition.
17. Nicholas Campion (2008) *The Dawn of Astrology*, Continuum Books, p.88.
18. John Michell *ibid.* p.26.
19. Clement of Alexandria *Stromata VI.4.35-7* quoted in J. Gardner Wilkinson *The Ancient Egyptians – Their Life and Customs* Vol I. Crescent Books, 1988.
20. The Hermetica that include the *Corpus Hermeticum,* are a body of writings believed to have originated in the ancient Egyptian temple libraries. They contain Egyptian occult teachings that have been contaminated while being translated and edited by Greek, Roman, Gnostic and Christian scholars. Lost to the West after classical times they survived in Byzantine and Arab libraries until the Renaissance. In 1945 a collection of Hermetic texts was discovered at Nag Hammadi in Egypt not far from Luxor.
21. Garth Fowden *The Egyptian Hermes,* p.76.
22. Joseph Campbell, quoted by Le Grice, *ibid.* p.17.
23. Could the name 'the Milky Way' derive from the fact that Nut was also represented in Egypt as 'the heavenly cow' who nourishes us all with her milk?
24. Herodotus *ibid.* p.158.
25. Paul Ghalioungui *Magic and Medical Science in Ancient Egypt,* p.47.
26. *The Emerald Tablet* is a short and cryptic piece of Hermetica that alchemists believe contains the secret of the *prima materia* and its transmutation. Although Hermes Trismegistus is the author named in the text, some doubt its antiquity, the first known appearance of the Emerald Tablet being in a book written in Arabic between the sixth and eighth centuries.
27. *Emerald Tablet* translated by Isaac Newton in his *Alchemical Papers,* Kings College Library Cambridge.
28. Herodotus *ibid.* pp.158–9.
29. Dennis Elwell *Cosmic Loom: The New Science of Astrology,* p.116.
30. Plotinus *Enneads* trans. A.H.Armstrong IV, iv, p.33.

Chapter 3

1. Right brain/left brain: the brain has two hemispheres which have different functions. The speech function, for example, resides in the left

hemisphere while the right hemisphere is better at recognising shapes and patterns. The left is involved in logic and reasoning and the right in intuitive understanding. The left is analytic and the right synthetic in the sense of being able to grasp things as wholes. Note that the left side of the body is controlled by the right hemisphere and vice versa.

2. Plato *Timaeus* trans. Benjamin Jowett www.gutenberg.org/ebooks/1572.
3. Richard Heath (2002) *The Matrix of Creation,* Bluestone Press.
4. Plato *Epinomis* quoted by Michell *ibid.* p.109.
5. Plato *The Laws* quoted by Michell *ibid.* pp.78–9.
6. Alexander Lauterwasser (2006) *Water Sound Images,* Macromedia Publishing.
7. Johannes Kepler *Harmonies of the World,* p.170.
8. Kepler *ibid.* p.209.
9. Kepler *ibid.* p.172
10. Subtle bodies are our non-physical bodies. According to esoteric teaching we each have a set of at least seven bodies – the physical, etheric, astral, emotional, mental, karmic and spiritual bodies (see p.117).
11. R. Clark (2000) p.98.
12. Malcolm Stewart *Patterns of Eternity,* p.179.
13. Revelation 22: The 12 precious stones set in the foundations of the Holy City within its circular wall correspond to the signs of the zodiac, indicating that it is designed according to the zodiac matrix.
14. See Phoebe Wyss *Hercules' Labours: the Evolutionary Path round the Zodiac.* An astrology student needs to be acquainted with the different archetypal qualities of the twelve phases of time cycles. This book, in which the myths of Hercules' labours are each allocated to one of the twelve houses of the horoscope, are used as keys to illuminate twelve areas of challenge in modern-life.
15. Herodotus *ibid.* pp.130–31.
16. See for example Lucia Gahlin *The Myths and Mythology of Ancient Egypt,* pp.50–57.
17. R. Clark, (2000), pp.54–5.
18. Gahlin *ibid.* p.50.
19. R. Clark *ibid.* p.64.
20. R. Clark, *ibid.* pp 80–81. (Djehuti is the Egyptian name for Thoth.)
21. See John Anthony West & Jan Gerhard Toonder *ibid.* p.198.

Chapter 4

1. R. Clark, (2000) p.17.
2. R. Schwaller de Lubicz *Sacred Science: the King of PharaonicTheocracy,* p.164.
3. John Anthony West *Serpent in the Sky: the High Wisdom of Ancient Egypt,* pp.156–61.

4. Herodotus *ibid.* p.130.
5. Herodotus *ibid.* p.147.
6. Herodotus *ibid.* p.150.
7. Herodotus *ibid.* p.159.
8. R.Clark *ibid.* pp.151–4.
9. Le Grice *ibid.* p.158.
10. Le Grice *ibid.* p.161.
11. C.G. Jung *The Freud-Jung Letters,* Letter 254.
12. Jung *Spirit in Man, Art and Literature,* p.56.
13. Liz Greene *Relating* p.62.
14. James Hillman, quoted by Tarnas *ibid.* p.83.
15. Richard Tarnas *The Passion of the Western Mind: Understanding the Ideas that have Shaped our World View.*
16. Tarnas *Cosmos and Psyche ibid.* p.67.
17. Tarnas *Cosmos and Psyche ibid.* p.161.
18. The California Institute of Integral Studies, San Francisco.
19. Teilhard de Chardin *The Phenomenon of Man,* p.56.

Chapter 5

1. Iamblichus *Mysteries* VIII. 4. 265 quoted by Fowden *ibid.* p.138.
2. R. Clark (2000), pp.44–6.
3. Michell *ibid,* pp.78–9.
4. Rig Veda I, 164, 1, quoted by Jane MacIntosh *A Peaceful Realm: The Rise and Fall of the Indus Civilisation*, p.99.
5. Plato *The Laws,* Book V.
6. R.T.Rundle Clark *Myth and Symbol in Ancient Egypt,* p.252.
7. R. Clark *ibid.* p.76.

Chapter 6

1. Knight & Lomas *ibid.*
2. *The Book of Enoch* trans R.H.Charles 1917, www.sacred-texts.com, LXXV, p.4.
3. *The Book of Enoch ibid.* LXXII, p.3.
4. Keith Critchlow *Time Stands Still* (The Gordon Fraser Gallery Ltd, 1979)
5. Michell *ibid.* p.153.
6. Michell *ibid.* p.81.
7. Lewis-Williams and Pearce *Inside the Neolithic Mind* (Thames & Hudson, 2005).
8. Knight and Lomas *ibid.* p.325.
9. Have a look on the internet at photos of the illumination of the Newgrange passage-tomb by the sun at sunrise on the winter solstice.
10. R. Clark (2000), p.221.

11. Rosemary Clark *The Sacred Magic of Ancient Egypt* (Llewellyn, 2003) pp.74–77. There is however no consensus on which son of Horus rules which direction. Jung favoured a different scheme of allocation.
12. *The Book of Enoch ibid.* LXXVI, p.103.
13. Ezekiel 1:10.

Chapter 7

1. MacIntosh *ibid.* p.99.
2. Joseph Campbell, *Creative Mythology: The Masks of God* Vol. 3, p.4.
3. See Rupert Sheldrake *A New Science of Life.*
4. In my view basic for students of astrology is to acquire an in-depth knowledge of the qualities and subject matter of the twelve archetypes and their fields of meaning. My book *Virtual Lives: the Animated Zodiac* was written for this purpose. Its aim is a non-fictional, didactic one, though its subject is presented in fictional form. Twelve characters representing the signs of the zodiac narrate their archetypical life stories in turn. These start in the 'interlife' before birth where they converse with their inner guide who prepares them for what lies ahead. In addition to their personalities, which are typical of their zodiac-signs, the events that happen to them, the situations in which they find themselves and the challenges they confront all conform to their archetypes.
5. See Nicolaus Klein & Rudiger Dahlke *Das Senkrechte Weltbild.*
6. The brain has two hemispheres which have different functions. The speech function, for example, resides in the left hemisphere while the right hemisphere is better at recognising shapes and patterns. The left is involved in logic and reasoning and the right in intuitive understanding. The left is analytic and the right synthetic in the sense of being able to grasp things as wholes. Note that the left side of the body is controlled by the right hemisphere and vice versa.
7. In modern astrology a number of different systems are in use for dividing the inner wheel of the houses into twelve sections. The 'whole sign house system' measures equal houses from 0 degrees of the rising sign. The equal house system also divides the inner wheel into twelve equal sections of 30 degrees starting at the degree of the ascendant. In this system the cusps of Houses 4 and 10 always make a 90-degree angle to the Ascendant, and the MC and the IC are entered separately as sensitive points. In all the other systems in use the houses are unequal as priority is given to the coincidence of the MC and IC points with the cusps of the tenth and fourth houses respectively. Twice in twenty-four hours, when 0 degrees Aries or 0 degrees Libra are rising on the Ascendant, the IC-MC axis will naturally lie at right-angles to the horizon, whereas at all other times it is oblique.

8. See Phoebe Wyss *Hercules' Labours: the Evolutionary Path round the Horoscope.*
9. Tarnas *Cosmos and Psyche, ibid.*
10. Origen *Homiliae in Leviticum* quoted by Jung *Collected Works ibid.* 16, 6
11. Have a look at an image of this beautiful mandala on the internet (search: Shri Yantra Mandala).

Chapter 8

1. As William Blake's chart was published during his lifetime, we presume he supplied the data himself, and the birth time must be more or less accurate.
2. Tarnas, *Cosmos and Psyche ibid.* pp.128–9.
3. In 1825 Blake's birth chart was published by R.C.Smith in the first and only edition of *Urania or The Astrologer's Chronicle and Mystical Magazine.*
4. Geoffrey Keynes, *Poetry and Prose of William Blake,* pp.638–9.
5. Ackroyd *Blake,* p.101.
6. Gilchrist *Life of William Blake,* p.341.
7. Ackroyd *ibid.* p.82.
8. Whitley, William T. *Art of England 1821-37,* p.34.
9. Ackroyd *ibid.* p.360.
10. They included the artists John Linnell, Samuel Palmer, Edward Calvert and Frederic Tatum, the group that went on to form 'The Ancients', an artists' colony based in Shoreham, Kent.
11. *The Literary Gazette* No 552 1827, p.541.
12. Gilchrist *ibid.* p.301.
13. Have a look at this painting on the internet (search: William Blake *Glad Day*).
14. Keynes *ibid.* pp.183–5.
15. Keynes *ibid.* p.375.
16. Keynes *ibid.* p.72.
17. Keynes *ibid.* pp.63–4.
18. Keynes *ibid.* p.75.
19. Ackroyd *ibid.* p.231.
20. Keynes *ibid.* p.98.
21. Ackroyd *ibid.* p.325.
22. Ackroyd *ibid.* p.232.
23. Have a look at these paintings on the internet (search: Blake, *Newton* and Blake *The Ancient of Days*).
24. Ackroyd *ibid.* p.194.
25. Ackroyd *ibid.* p.87.
26. Keynes *ibid.* p.873.

27. Keynes *ibid.* p.121.
28. Ackroyd *ibid.* p.325.
29. Keynes *ibid.* p.78.
30. Tarnas, *Cosmos and Psyche, ibid.* pp.143–4. The whole of Chapter 4 is valuable for understanding the Uranus-Pluto cycle historically and as a context for Blake's ideas.
31. Keynes *ibid.* pp.435–6.
32. Keynes *ibid.* p.118.
33. Keynes *ibid.* p.861.
34. Keynes *ibid.* p.187.
35. Keynes *ibid.* p.447.
36. Le Grice *ibid.* p.76.

Conclusion

1. Plato *The Republic.*
2. Le Grice *ibid.* p.19.
3. Ervin Laszlo *The Chaos Point: The World at the Crossroads,* p.5.
4. Anne Baring *The Dream of the Cosmos: a Quest for the Soul,* pp.xxiii–xxiv.
5. Plotinus *Enneads* trans. A.H.Armstrong IV, iii, p.9.
6. See for example Brian Swimme *The Hidden Heart of the Cosmos: Humanity and the New Story.*
7. Origen *Homiliae in Leviticum* in Jung *Collected Works* 16, p.197.
8. Heraclitus *Fragments.*
9. See Tarnas *Cosmos and Psyche, ibid.*
10. Practising meditation takes us deeper into ourselves, transforming our normal everyday consciousness so that our awareness can come from a place of silence and oneness. If we work with astrology from this place, we not only become open to the wider and deeper levels of archetypal meaning, but our astrology becomes an act of co-creation.
11. Elwell *ibid.*
12. Le Grice *ibid.* p.20.
13. Epigram attributed to Ptolemy quoted by Fowden, *ibid.* p.93.
14. The *Book of Enoch, ibid.* LXXI

Appendices

A. The astrological glyphs

♈	Aries	☉	Sun
♉	Taurus	☽	Moon
♊	Gemini	☿	Mercury
♋	Cancer	♀	Venus
♌	Leo	♂	Mars
♍	Virgo	♃	Jupiter
♎	Libra	♄	Saturn
♏	Scorpio	♅ ⛢	Uranus
♐	Sagittarius	♆	Neptune
♑	Capricorn	⯓ ♇	Pluto
♒	Aquarius		
♓	Pisces		

B. Table of Archetypal Correspondences

Sign	House	Element	Mode	Gender	Planet
Aries	1	Fire	Cardinal	Male	Mars
Taurus	2	Earth	Fixed	Female	Venus
Gemini	3	Air	Mutable	Male	Mercury
Cancer	4	Water	Cardinal	Female	Moon
Leo	5	Fire	Fixed	Male	Sun
Virgo	6	Earth	Mutable	Female	Mercury
Libra	7	Air	Cardinal	Male	Venus
Scorpio	8	Water	Fixed	Female	Pluto
Sagittarius	9	Fire	Mutable	Male	Jupiter
Capricorn	10	Earth	Cardinal	Female	Saturn
Aquarius	11	Air	Fixed	Male	Uranus
Pisces	12	Water	Mutable	Female	Neptune

C. Keywords for the Houses

D. Aspects referred to in this book

1. **Conjunction:** 0° aspect, 10° orb, either harmonious or discordant depending on the planets involved. Number = 1, Element = Fire/Sun
2. **Opposition:** 180° aspect, 10° orb, discordant. Number = 2, Element = Water/Moon.
3. **Trine:** 120° aspect, 10° orb, harmonious. Number = 3, Element = Fire/Jupiter.
4. **Square:** 90° aspect, 10° orb, discordant. Number = 4, Element = earth/Saturn.
5. **Quincunx:** 150° aspect, 3° orb, discordant. Number = 12, Element = water/Pluto.
6. **Sextile:** 60° aspect, 6° orb, harmonious. Number = 6. Element = air/Mercury.
7. **Quintile:** 72° aspect, 2° orb, harmonious. Number = 5.
8. **Semi-Sextile:** 30° aspect.

E. Zodiac archetypal field diagrams

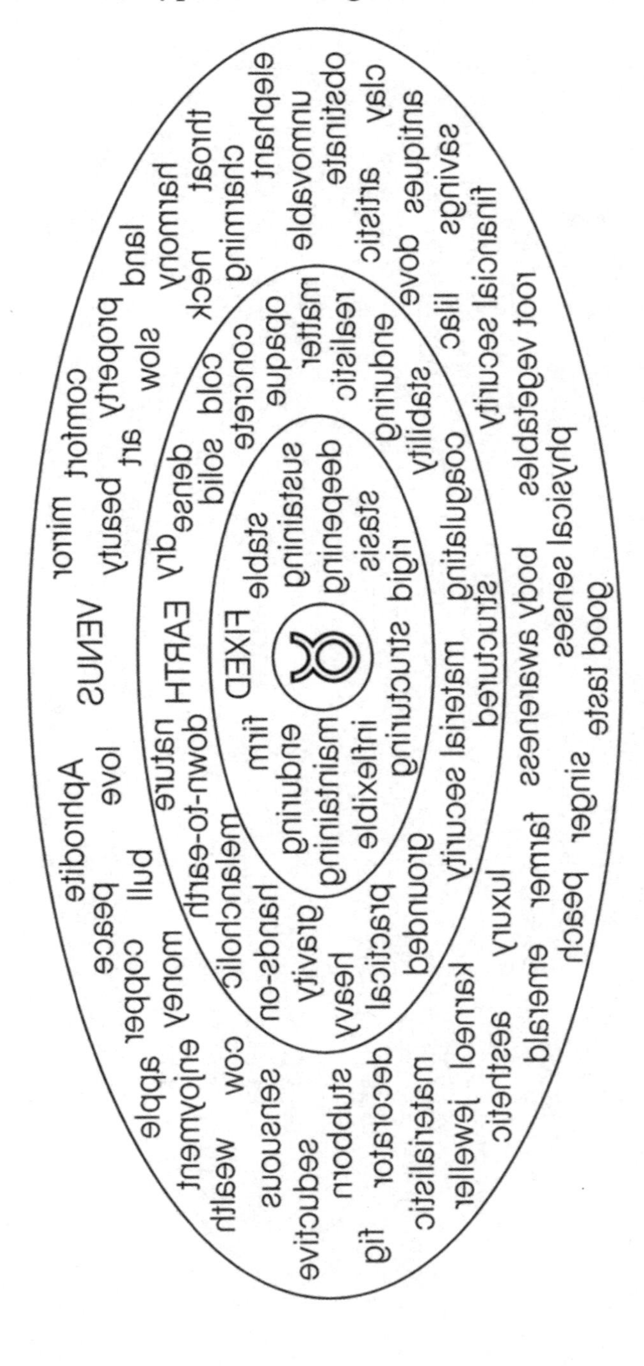

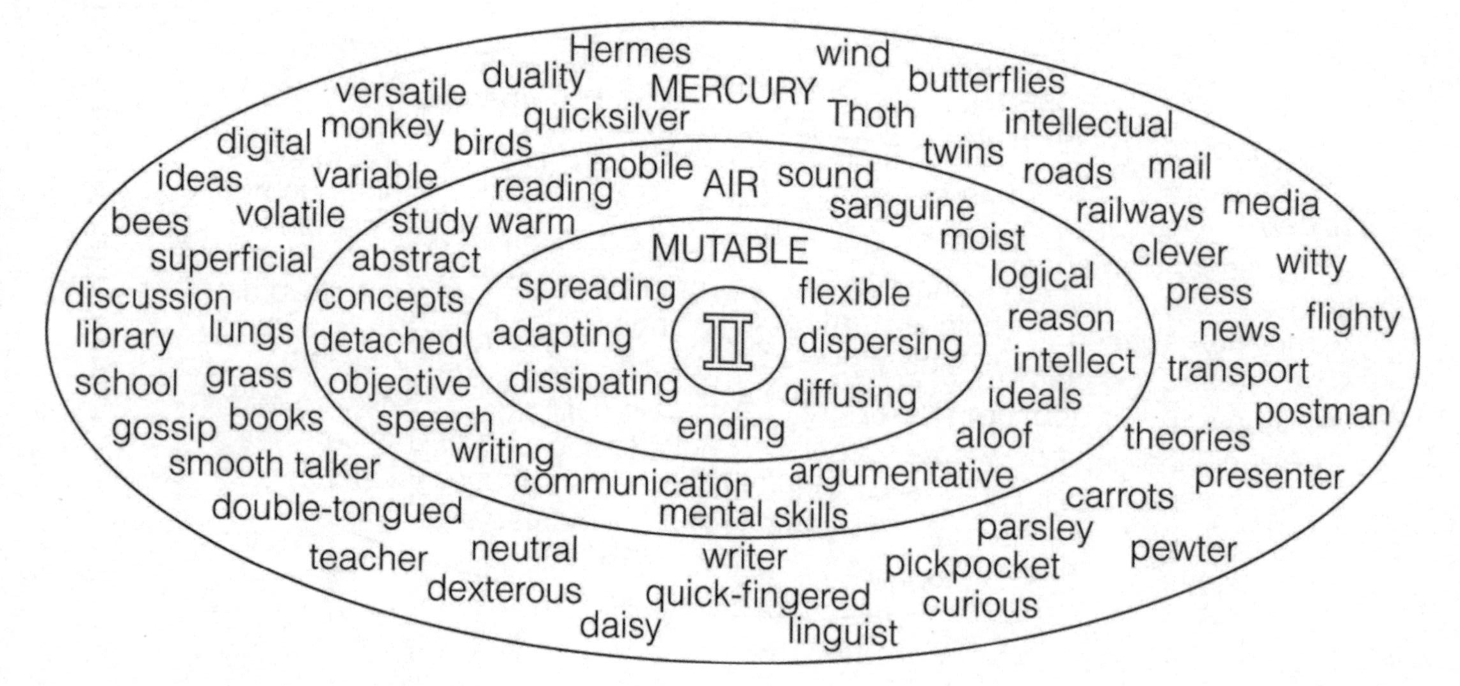
♊
MUTABLE
spreading
flexible
adapting
dispersing
dissipating
diffusing
ending
AIR
mobile
reading
sound
sanguine
study warm
moist
abstract
logical
concepts
reason
detached
intellect
objective
ideals
speech
aloof
writing
communication
argumentative
mental skills
Hermes
wind
duality
MERCURY
butterflies
versatile
quicksilver
Thoth
intellectual
digital
monkey
birds
twins
ideas
variable
roads
mail
bees
volatile
railways
media
superficial
clever
witty
discussion
press
flighty
library
lungs
news
school
grass
transport
gossip
books
postman
theories
smooth talker
presenter
double-tongued
carrots
parsley
teacher
neutral
writer
pickpocket
pewter
dexterous
quick-fingered
curious
daisy
linguist

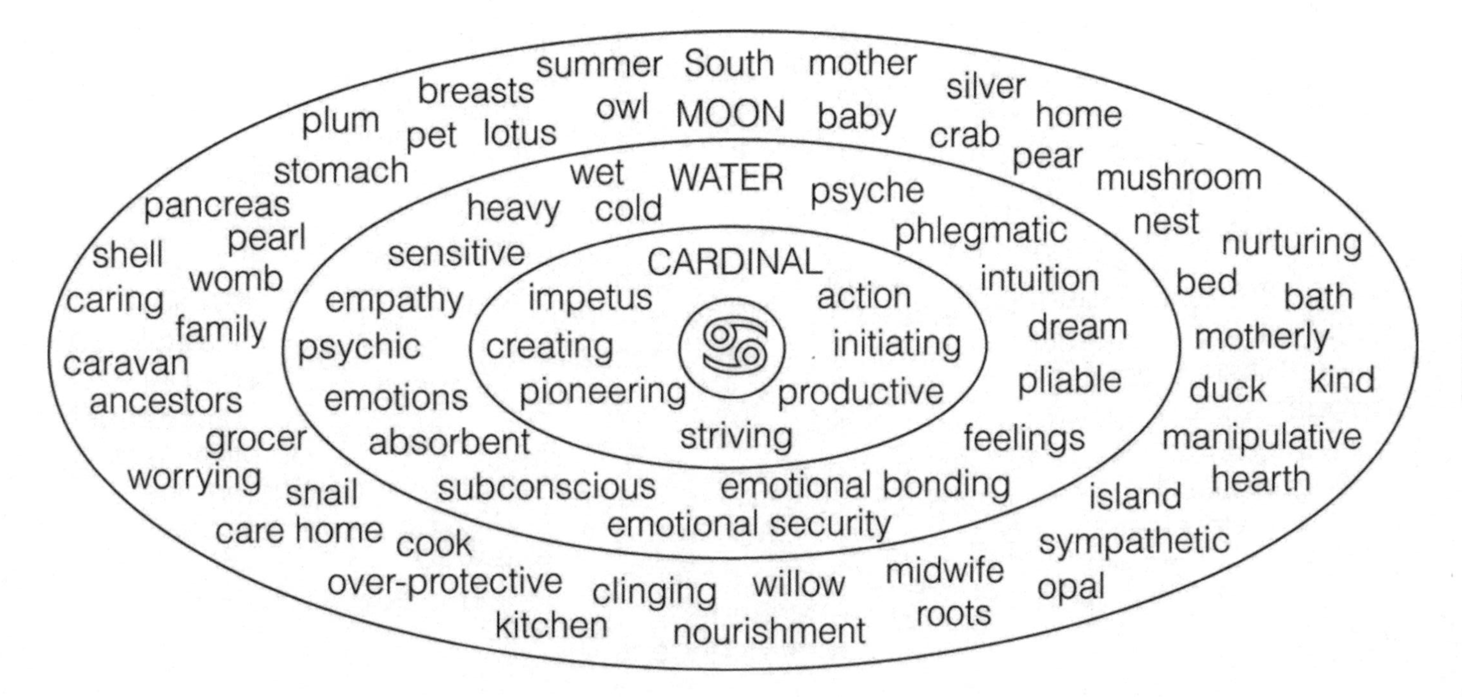

summer South mother
breasts owl MOON baby silver
plum pet lotus crab home
stomach pear mushroom
pancreas pearl
shell womb
caring family
caravan
ancestors
grocer
worrying snail
care home cook
over-protective clinging willow midwife
kitchen nourishment roots
opal
sympathetic
island hearth
manipulative
duck kind
motherly
bed bath
nest nurturing
wet WATER psyche
heavy cold
phlegmatic
sensitive
empathy
psychic
emotions
absorbent
subconscious
emotional security
emotional bonding
feelings
pliable
dream
intuition
CARDINAL
impetus
action
creating
initiating
pioneering
productive
striving

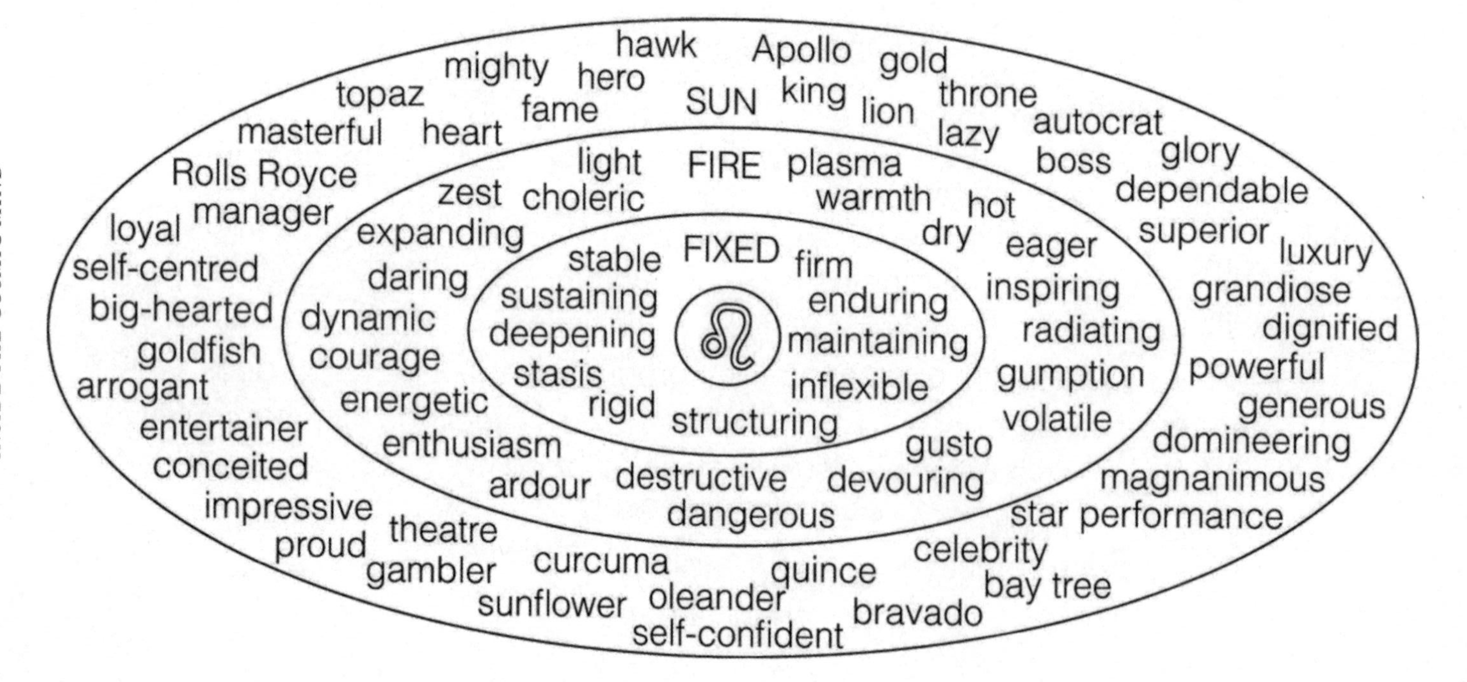
hawk
Apollo
gold
mighty
hero
SUN
king
throne
topaz
fame
lion
autocrat
masterful
heart
lazy
boss
glory
Rolls Royce
dependable
loyal manager
superior
luxury
self-centred
grandiose
big-hearted
dignified
goldfish
powerful
arrogant
generous
entertainer
domineering
conceited
magnanimous
impressive
star performance
proud
theatre
celebrity
gambler
curcuma
quince
bay tree
sunflower
oleander
bravado
self-confident
light
FIRE
plasma
zest
choleric
warmth
hot
expanding
dry
eager
daring
inspiring
dynamic
radiating
courage
gumption
energetic
volatile
enthusiasm
gusto
ardour
destructive
devouring
dangerous
stable
FIXED
firm
sustaining
enduring
deepening
maintaining
stasis
inflexible
rigid
structuring

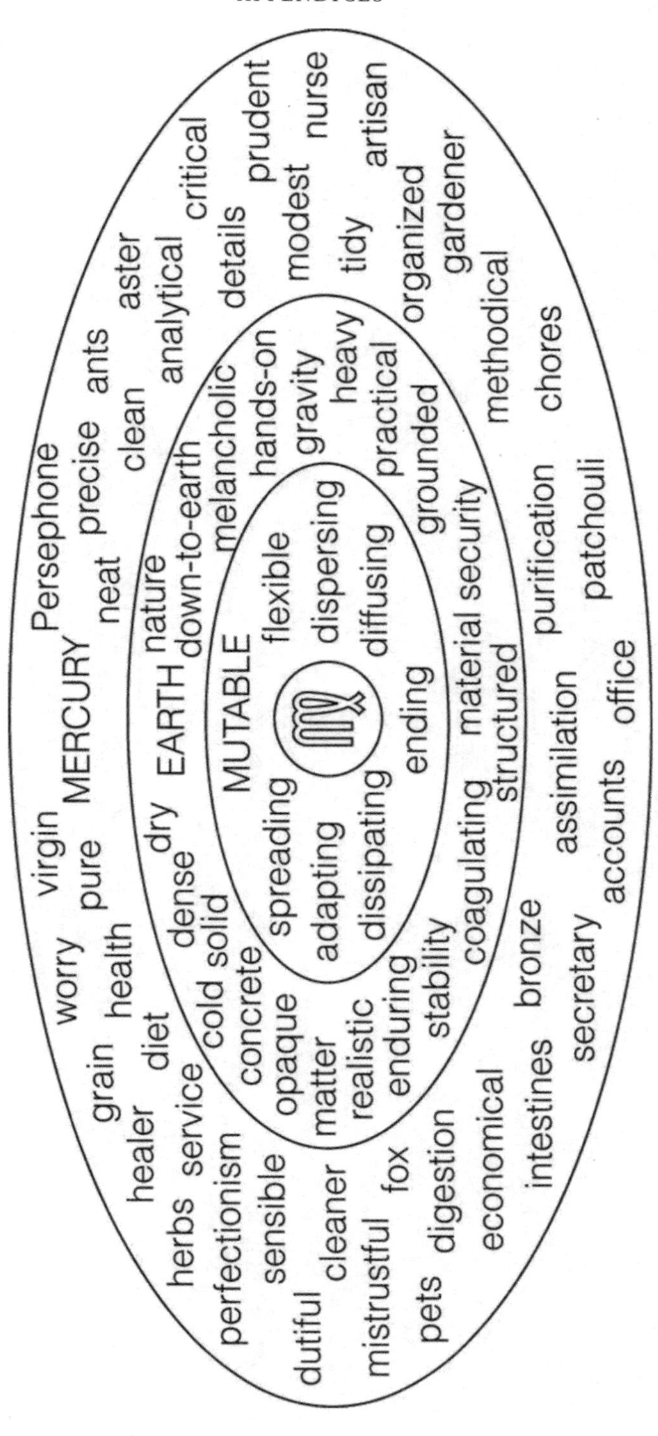
worry virgin MERCURY Persephone
grain health pure neat precise ants aster
healer diet clean analytical critical
herbs service details prudent
perfectionism modest nurse
sensible tidy artisan
dutiful cleaner organized gardener
mistrustful fox methodical
pets digestion chores
economical bronze assimilation purification
intestines secretary accounts office patchouli
EARTH nature down-to-earth
dry dense cold solid concrete
melancholic hands-on
opaque matter gravity heavy
realistic practical
enduring stability grounded
material security
coagulating structured
MUTABLE
spreading flexible
adapting dispersing
dissipating diffusing
ending

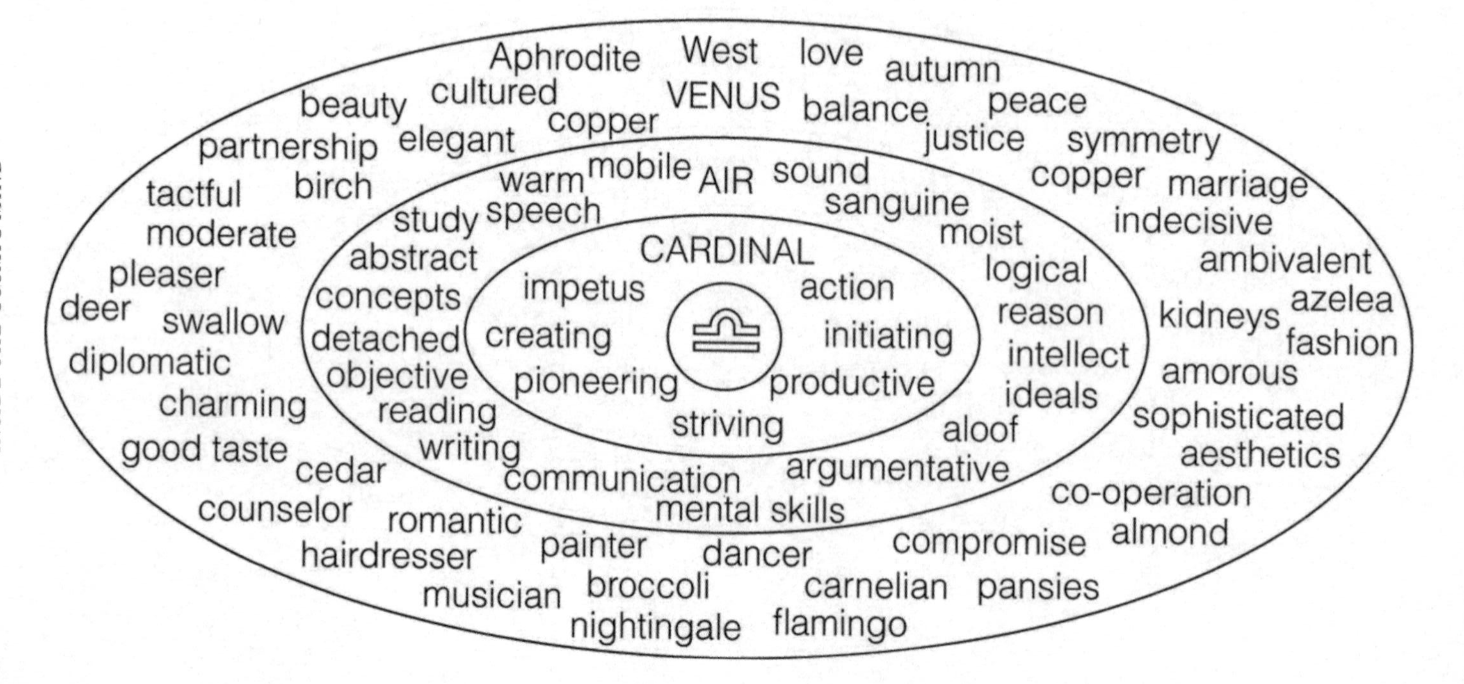
Aphrodite
West
love
autumn
beauty
cultured
VENUS
balance
peace
partnership
elegant
copper
justice
symmetry
tactful
birch
copper
marriage
moderate
indecisive
pleaser
ambivalent
deer
swallow
kidneys
azelea
diplomatic
fashion
charming
amorous
good taste
sophisticated
cedar
aesthetics
counselor
romantic
co-operation
hairdresser
painter
dancer
compromise
almond
musician
broccoli
carnelian
pansies
nightingale
flamingo
warm
mobile
AIR
sound
study
speech
sanguine
moist
abstract
logical
concepts
reason
detached
intellect
objective
ideals
reading
aloof
writing
communication
argumentative
mental skills
CARDINAL
impetus
action
creating
initiating
pioneering
productive
striving

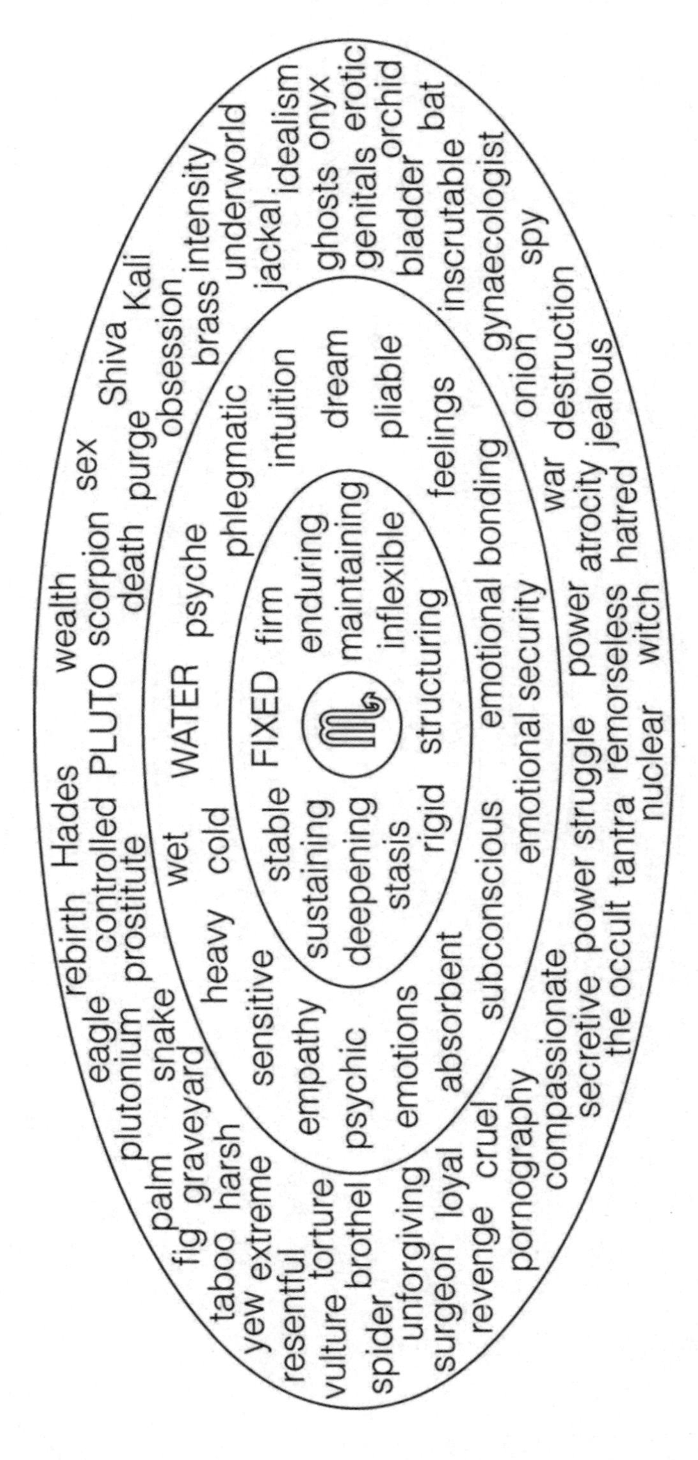
fig palm taboo yew extreme harsh graveyard plutonium eagle rebirth Hades
controlled prostitute snake PLUTO wealth scorpion death sex purge Shiva Kali
obsession brass intensity underworld jackal idealism ghosts onyx genitals erotic orchid
bladder bat inscrutable gynaecologist spy onion destruction jealous war atrocity hatred
power remorseless witch nuclear power struggle tantra the occult secretive compassionate
pornography cruel loyal revenge surgeon unforgiving spider brothel torture vulture resentful
WATER wet cold heavy sensitive empathy psychic emotions absorbent subconscious
emotional security emotional bonding feelings pliable dream intuition phlegmatic psyche
FIXED firm enduring maintaining inflexible structuring rigid stasis deepening sustaining stable

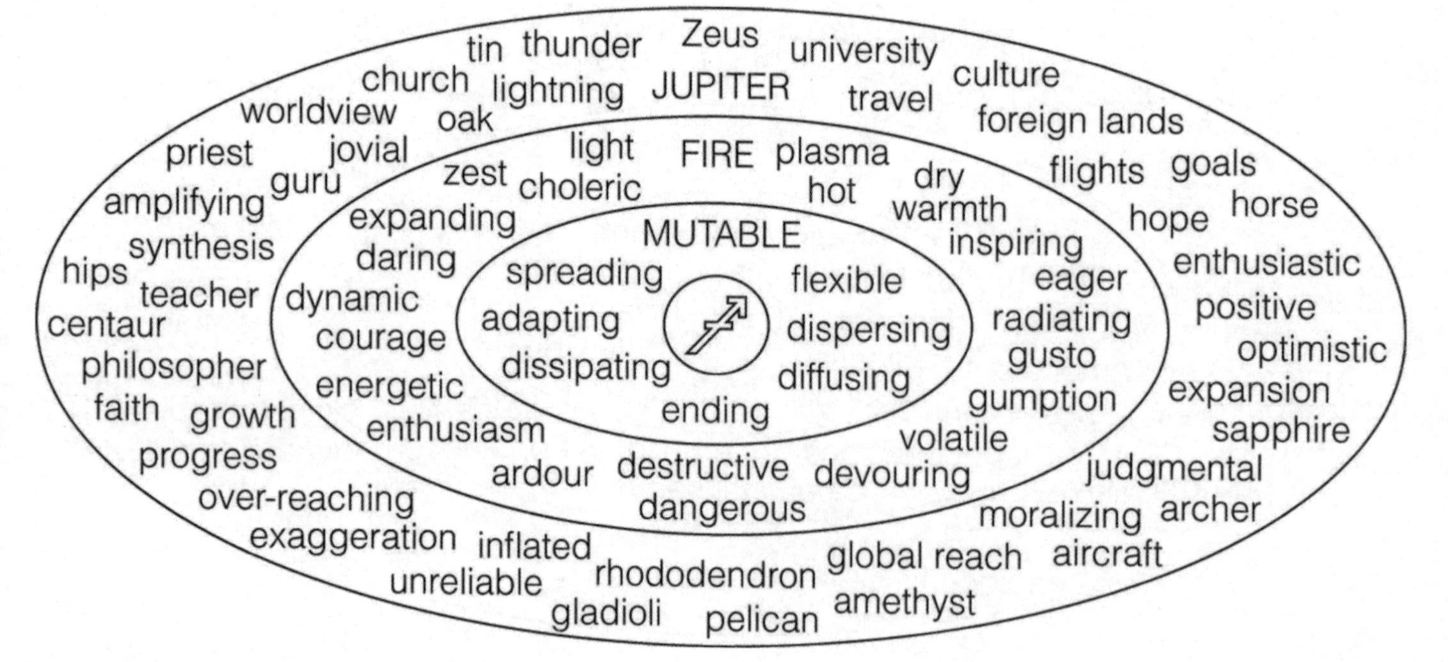
tin thunder Zeus university
church lightning JUPITER travel culture
worldview oak foreign lands
priest jovial light FIRE plasma
amplifying guru zest choleric hot dry flights goals
expanding warmth hope horse
synthesis MUTABLE inspiring
hips daring spreading flexible eager enthusiastic
teacher dynamic radiating positive
centaur adapting dispersing
courage gusto optimistic
philosopher dissipating diffusing
energetic gumption expansion
faith growth ending
enthusiasm volatile sapphire
progress ardour destructive devouring judgmental
over-reaching dangerous moralizing archer
exaggeration inflated global reach aircraft
unreliable rhododendron amethyst
gladioli pelican

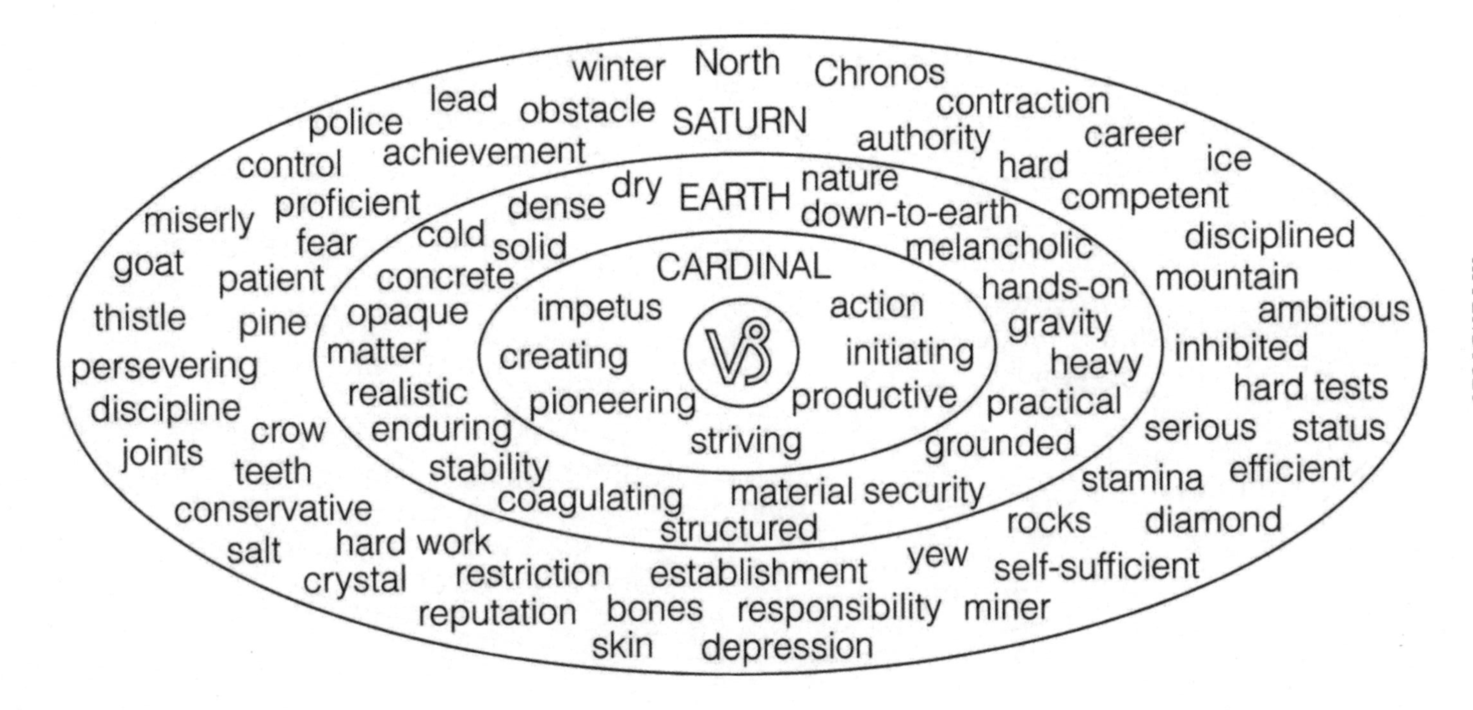
winter
North
Chronos
lead
obstacle
SATURN
contraction
police
achievement
authority
career
control
hard
ice
miserly
proficient
competent
fear
disciplined
goat
patient
mountain
thistle
pine
ambitious
persevering
inhibited
discipline
crow
hard tests
joints
teeth
serious
status
conservative
stamina
efficient
salt
hard work
rocks
diamond
crystal
restriction
establishment
yew
self-sufficient
reputation
bones
responsibility
miner
skin
depression
EARTH
dry
dense
nature
down-to-earth
cold
solid
melancholic
concrete
hands-on
opaque
gravity
matter
heavy
realistic
practical
enduring
grounded
stability
coagulating
material security
structured
CARDINAL
impetus
action
creating
initiating
pioneering
productive
striving
♑

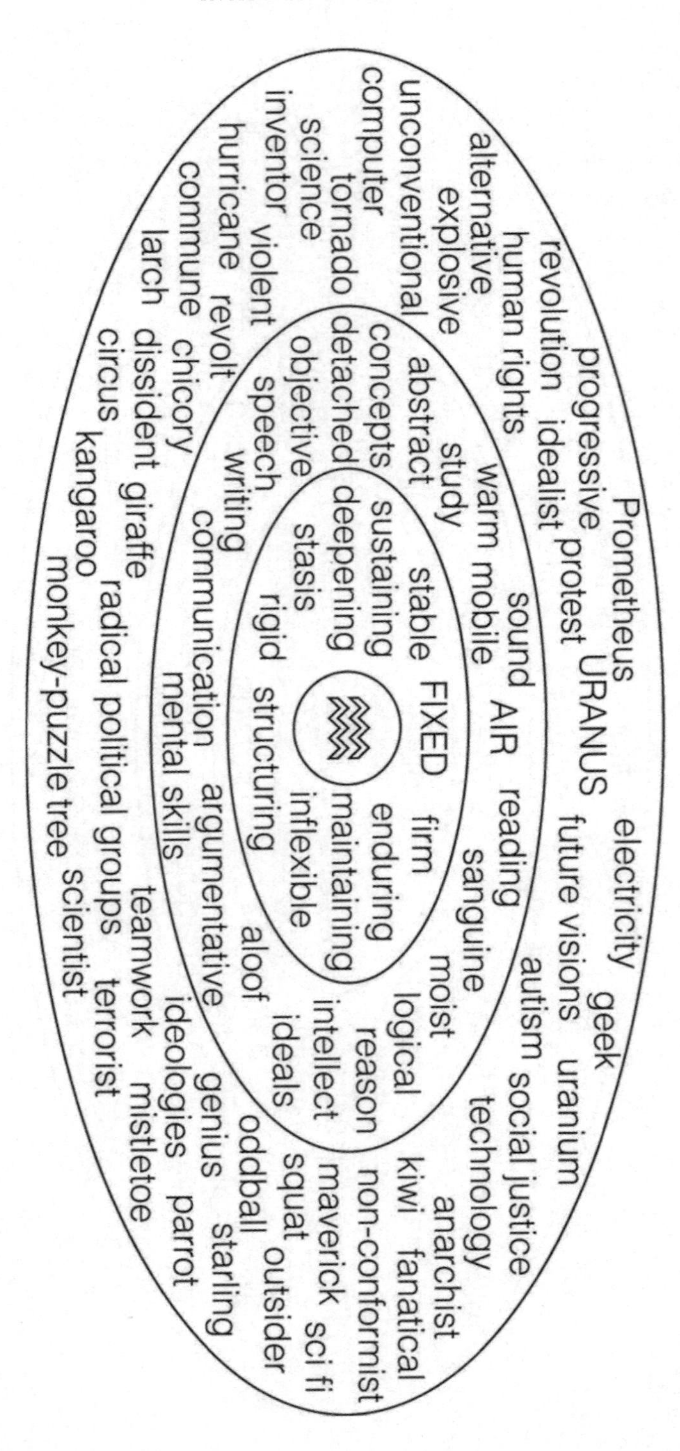
Prometheus
electricity
geek
progressive
URANUS
future visions
uranium
revolution
idealist
protest
autism
social justice
human rights
technology
alternative
anarchist
explosive
unconventional
computer
tornado
science
inventor
violent
hurricane
revolt
commune
chicory
larch
dissident
giraffe
circus
kangaroo
radical political groups
monkey-puzzle tree
scientist
teamwork
terrorist
mistletoe
parrot
ideologies
genius
starling
oddball
outsider
squat
maverick
sci fi
non-conformist
kiwi
fanatical
sound
AIR
reading
warm
mobile
sanguine
moist
study
logical
abstract
reason
concepts
intellect
detached
objective
ideals
speech
writing
communication
argumentative
mental skills
aloof
stable
FIXED
firm
sustaining
enduring
deepening
maintaining
stasis
inflexible
rigid
structuring
♒

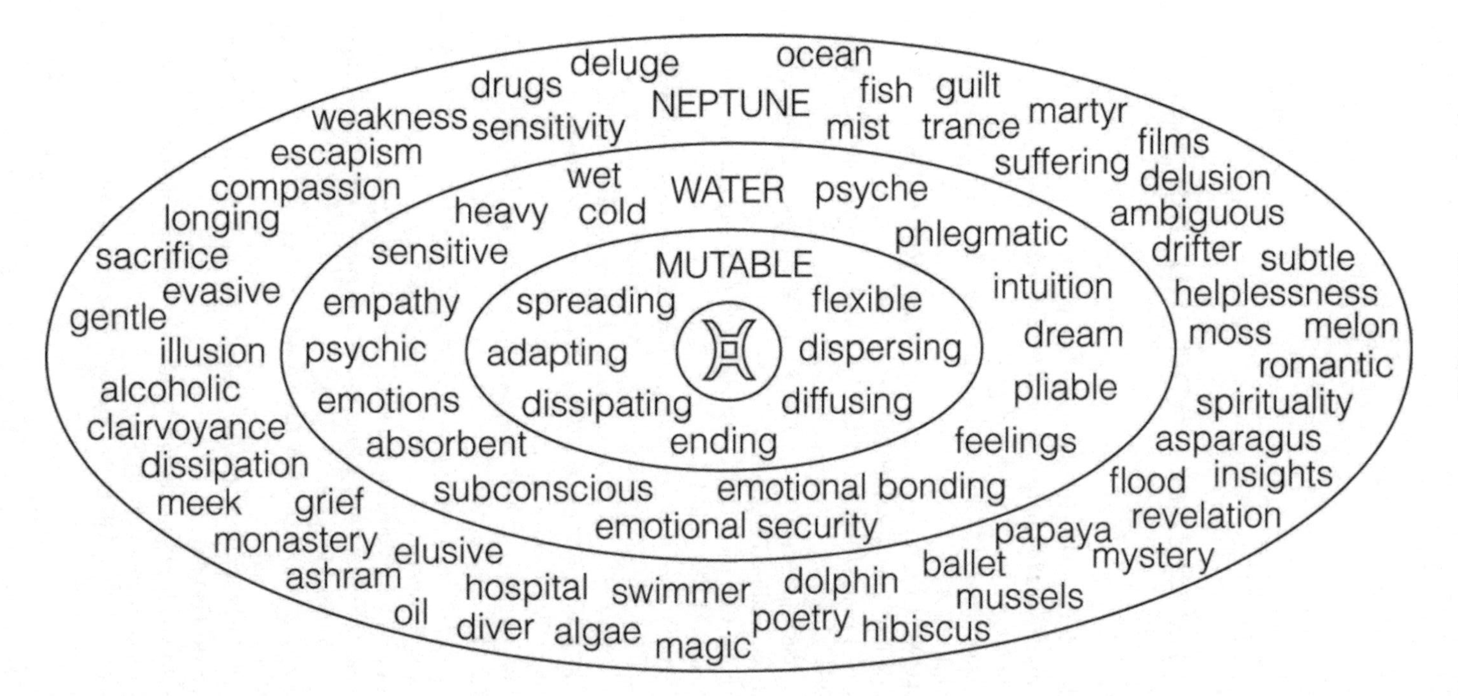
ocean
deluge
drugs
NEPTUNE
fish
guilt
weakness
sensitivity
mist
trance
martyr
escapism
films
suffering
delusion
compassion
ambiguous
longing
WATER
wet
cold
heavy
psyche
phlegmatic
sacrifice
sensitive
drifter
subtle
MUTABLE
gentle
evasive
empathy
spreading
flexible
intuition
helplessness
illusion
psychic
adapting
dispersing
dream
moss
melon
romantic
alcoholic
emotions
dissipating
diffusing
pliable
spirituality
clairvoyance
absorbent
ending
feelings
asparagus
dissipation
subconscious
emotional bonding
flood
insights
meek
grief
emotional security
revelation
monastery
elusive
papaya
mystery
ashram
ballet
hospital
swimmer
dolphin
mussels
oil
diver
algae
magic
poetry
hibiscus

References And Further Reading

Ackroyd, Peter (1995). *Blake,* London: QPD.

Baring, Anne (2013) *The Dream of the Cosmos: a Quest for the Soul,* Wimborne, Dorset: Archive Publications.

Bauval, Robert & Gilbert Adrian (1994) *The Orion Mystery,* London/New York: *BCA.*

Blake, William *Poetry and Prose of William Blake* ed. Geoffrey Keynes (1961) London: The Nonesuch Library.

Bohm, David (1980/2002) *Wholeness and the Implicate Order.* Rep. London: Routledge.

Campbell, Joseph (1976/1991) *Creative Mythology: The Masks of God,* Vol. 3. Rep. London: Penguin.

Nicholas Campion (2012) *Astrology and Cosmology in the World's Religions,* New York: New York University Press.

—, (2008) *The Dawn of Astrology,* London/New York Continuum Books.

Capra, Fritjof (1982) *The Turning Point,* Simon & Shuster.

—, (1975/1983) *The Tao of Physics,* Glasgow: William Collins.

Clark, Glenn (2007) *The Man who Tapped the Secrets of the Universe,* Filiquarian Publishing, LLC.

Clark, Rosemary (2000) *The Sacred Tradition in Ancient Egypt,* Woodbury MN: Llewellyn.

—, (2003) *The Sacred Magic of Ancient Egypt,* Woodbury MN: Llewellyn.

Critchlow Keith (1979/2007) *Time Stands Still,* London: Gordon Fraser Gallery. Rep. Edinburgh, Floris Books.

Devereux, Paul (2001) *Stone Age Sound Tracks: the Acoustic Archaeology of Ancient Sites,* London: Vega.

Dobyns, Zipporah (1983) *Expanding Astrology's Universe,* San Diego CA: ACS Publications.

Elwell, Dennis (1987) *Cosmic Loom: The New Science of Astrology,* London: Unwin Hyman.

Wilkinson, J.Gardner (1988/1990) *The Ancient Egyptians – Their Life and Customs,* London: John Murray.

Fowden, Garth (1986) *The Egyptian Hermes: A Historical Approach to the Late*

Pagan Mind, Cambridge: CUP.
Gahlin, Lucia (2003) *The Myths and Mythology of Ancient Egypt* London: Anness Publishing Ltd.
—, (2002) *Egyptian Religion,* London: Anness Publishing Ltd.
Ghalioungui, Paul (1963) *Magic and Medical Science in Ancient Egypt,* Hodder & Stoughton.
Gilchrist, Alexander (1863) *Life of William Blake.*
Greene, Liz (1977/1983) *Relating,* London: Coventure Ltd.
Hancock, Graham & Santa Faiia (1998) *Heaven's Mirror:Quest for the Lost Civilization,* New York: Crown Publishers.
Hope, Murry (1996) *The Sirius Connection:Unlocking the Secrets of Ancient Egypt,* Shaftesbury, Dorset: Element Books.
Heath, Richard (2002) T*he Matrix of Creation: Technology of the Gods,* St Dogmaels: Bluestone Press.
Herodotus *The Histories,* trans. A de Selincourt (1954) Harmondsworth, Middlesex: Penguin.
Hillman, James (1983) *Archetypal Psychology: a Brief Account,* Putnam: CT: Spring Publications.
Hyde, Maggie (1992) *Jung and Astrology,* London: The Aquarian Press.
Jung, Carl Gustav (1955/1977) *Synchronicity: an Acausal Connecting Principle,* trans. R.F.C Hull, Rep. London: Routledge & Kegan Paul.
—, (1953-1979) *The Collected Works of C.G.Jung,* 19 vols trans. R.F.C Hull Princeton: Princeton University Press; London Routledge & Kegan Paul.
Kepler, Johannes (1995) *Epitome of Copernican Astronomy & Harmonies of the World,* New York: Prometheus Books.
Keynes, Geoffrey (1961) *Poetry and Prose of William Blake,* London: The Nonesuch Library.
Klein Nicolaus & Dahlke Rudiger (1986) *Das Senkrechte Weltbild,* Munich: Hugendubel.
Knight, Christopher & Lomas, Robert (2000) *Uriel's Machine: the Ancient Origins of Science,* London: Arrow Books.
Lamy, Lucie (1981) *Egyptian Mysteries,* Inner Traditions International.
Laszlo, Ervin (2006) *The Chaos Point: The World at the Crossroads,* London: Piatkus Books.
Lauterwasser, Alexander (2006) *Water Sound Images,* Macromedia Publishing.
Le Grice, Keiron (2010) *The Archetypal Cosmos: Rediscovering the Gods in Myth, Science and Astrology,* Edinburgh: Floris Books.
—, (2012) *Discovering Eris,* Edinburgh: Floris Books.
Lewis-Williams, David and Pearce, David (2005/2009) *Inside the Neolithic Mind,* London: Thames & Hudson.
MacIntosh, Jane (2002) *A Peaceful Realm: The Rise and Fall of the Indus Civilization,* Nevraumont Publishing Co. Inc.
Michell, John (1973) *City of Revelation,* London: Abacus.

—,(1988) *The Dimensions of Paradise: The Proportions and Symbolic Numbers of Ancient Cosmology,* New York: Harper & Row.

Peat, David (1987) *Synchronicity: the Bridge between Matter and Mind,* New York: Bantam Books.

Plato (1977) Timaeus and Critias. Trans. Desmond Lee. London: Penguin Books.

—, *The Laws* trans. A.E.Taylor, in *The Collected Dialogues,* ed. Edith Hamilton & Huntington Cairns, Princeton, N.J: Princeton University Press.

—, *The Republic The Internet Classics Archive, trans.* B.Jowett.

Plotinus *Enneads* Trans. A.H.Armstrong (1929) Harvard University Press.

Rundle Clark, R.T. (1959/1978) *Myth and Symbol in Ancient Egypt,* London: Thames & Hudson.

Santillana, G. & von Dechend, H. (1969) *Hamlet's Mill,* USA: Gambit.

Schwaller de Lubicz, R.A.(1988) *Sacred Science : the King of Pharaonic Theocracy,* Inner Traditions International.

—, (1977) *The Temple Within Man,* trans. Robert and Deborah Lawlor USA: Autumn Books.

Sheldrake, Rupert (1981/1995 *A New Science of Life: the Hypothesis of Morphic Resonance,* Rep. Rochester, VT: Park Street Press.

—, (1988) *The Presence of the Past,* London: Collins.

Skinner, Stephen (2006) *Sacred Geometry: Deciphering the Code,* London: Gaia Books.

Stewart, Malcolm (2009) *Patterns of Eternity,* Edinburgh: Floris Books.

Swimme, Brian (1996) *The Hidden Heart of the Cosmos: Humanity and the New Story,* Maryknoll, NY: Orbis Books.

Tarnas, Richard (1991/1993) *The Passion of the Western Mind: Understanding the Ideas that have Shaped our World View,* Rep. London:Pimlico.

—, (2006) *Cosmos and Psyche: Intimations of a New World View,* New York: Viking.

Taylor, Ken (2012) *Celestial Geometry: Understanding the Astronomical Meanings of Ancient Sites,* London: Watkins.

Teilhard de Chardin, Pierre (1965) *The Phenomenon of Man.* New York: Harper and Row.

West, John Anthony (1993) *Serpent in the Sky: the High Wisdom of Ancient Egypt,* Wheaton IL: Quest Books.

West, John Anthony & Toonder, Jan Gerhard (1970/1973) *The Case for Astrology,* Harmondsworth Middlesex: Penguin Books.

Whitley, William T. (1930) *Art of England 1821-37.*

Woolger, Roger (1988) *Other Lives Other Selves,* Doubleday.

Wyss, Phoebe (2007) *Hercules Labours: the Evolutionary Path round the Horoscope,* Crediton Devon: Treetongue.

—, (2008) *Virtual Lives: the Animated Zodiac,* Crediton Devon: Treetongue.

Index

The Archetypal Cosmos

Rediscovering the Gods in Myth, Science and Astrology

Keiron Le Grice

Le Grice has a gift, perhaps even a genius, for extremely clear assessments, expositions, and formulations of complex ideas – all grounded in a deeper vision of things, which makes this clarity possible.

— Richard Tarnas, author of *The Passion of the Western Mind* and *Cosmos and Psyche*

The modern world is passing through a time of critical change on many levels: cultural, political, ecological and spiritual. We are witnessing the decline and dissolution of the old order, the tumult and uncertainty of a new birth. Against this background, there is an urgent need for a coherent framework of meaning to lead us beyond the growing fragmentation of culture, belief and personal identity.

Keiron Le Grice argues that the developing insights of a new cosmology could provide this framework, helping us to discover an underlying order shaping our life experiences. He draws especially on the work of C. G. Jung, Joseph Campbell, Richard Tarnas, Fritjof Capra, David Bohm and Brian Swimme.

Heralding a 'rediscovery of the gods' and the passage into a new spiritual era, *The Archetypal Cosmos* presents a new understanding of the role of myth and archetypal principles in our lives, one that could give a cosmic perspective and deeper meaning to our personal experiences.

florisbooks.co.uk